EVALUATION OF INSECTICIDE RESISTANCE MANAGEMENT PROGRAMME

— Theory and Practice —

EVALUATION OF INSECTICIDE RESISTANCE MANAGEMENT PROGRAMME
— Theory and Practice —

Rajinder Peshin
Associate Professor
Division of Agricultural Extension Education
FOA, Sher-e- Kashmir University of Agricultural Sciences and Technology of Jammu
Chatha, Jammu – 180 009

2009

DAYA PUBLISHING HOUSE

Delhi - 110 035

ISBN 81-7035-605-9
ISBN 978-81-7035-605-9

Published by : **Daya Publishing House**
1123/74, Deva Ram Park
Tri Nagar, Delhi - 110 035
Phone: (011) 27383999
Fax: (011) 23260116
e-mail : dayabooks@vsnl.com
website : www.dayabooks.com

Showroom : 4760-61/23, Ansari Road, Darya Ganj,
New Delhi - 110 002
Phone: 23245578, 23244987

Laser Typesetting : **Classic Computer Services**
Delhi - 110 035

Printed at : **Chawla Offset Printers**
Delhi - 110 052

PRINTED IN INDIA

To my teachers

Late G. N. Peshin
T. N. Misri
Prof. A. J. S. Dhaliwal
Prof. Raghbir Singh

PREFACE

Use of pesticides was intensified in Indian agriculture with the advent of the green revolution in mid 1960s, well after the release of the book *"Silent Spring"* authored by Rachel Carson in 1962, which highlighted the negative externalities of pesticide use. It is an irony that when scientific community in the developed world accepted the concept of integrated control, Indian farmers were exposed to technological paradigm of calendar based pesticide use as a solution for all pest related problems. The losses caused by pests more than doubled after the introduction of synthetic pesticides. In case of cotton crop, it almost tripled and led farmers being caught into 'pesticide treadmill'.

To combat the negative externalities of pesticide use, first integrated pest management (IPM) project was launched in India, in 1974-75 under Operational Research Project. Since then different IPM programmes have been implemented. IPM activities were intensified in 1990s. However, the worth and merit of these programmes at farmers' level is almost unknown and information available is sketchy. The book "Evaluation of Insecticide Resistance Management Programme-Theory and Practice" is a small step to document the worth of IPM programme.

The book deals with concepts and methodologies for evaluation of agricultural extension programmes in general and IPM programmes in particular. The aim of the book is to highlight the role of evaluation research based on theory and practice of evaluation. In India, no attention has been paid to evaluation of agricultural programmes. In the developed world, like the United States of America, programme evaluation is inbuilt with all development programmes and the universities have separate departments/sections dealing with programme evaluation. The evaluation of IPM programmes based on the evaluation methodologies provides feedback to stakeholders.

Since 2002, insecticide resistance management based IPM programme in cotton is implemented in 10 cotton growing states with a funding of US$ 2.58 million from the Ministry of Agriculture, Government of India under the Technology Mission on Cotton (TMC), Mini-Mission (MM-II) programme. The worth of this programme in terms of the evaluation indicators at farmers' level *viz.* gain in knowledge and analytical skills; adoption, adoptability and adaptability; and reduction in pesticide use are almost unknown. The project reports complied by the implementing agencies indicate the gain in terms of reduction in pesticide expenditure and increased productivity as has been the case with other IPM programmes implemented right from Operational Research Project. The book documents the worth of IRM programme in Punjab based on evaluation methodologies.

First chapter provides an overview of the IRM. Second chapter deals with development, implementation and impact of IPM programmes around the world. The concepts and methodologies of evaluation are detailed in chapter three. Chapter four deals with the methods adopted for evaluation of IRM programme in Punjab. Fifth chapter presents the evaluation results of the IRM programme. The last chapter summarizes the evaluative study and draws conclusion.

The book intends to generate the interest of policy makers in Indain agriculture towards evaluation research. The book is a contribution of the discipline of extension education to the inter-disciplinary knowledge relating to IPM programmes. It will serve the agencies associated with IPM programmes/projects in formulating IPM programmes to fit the farmers. At post-graduate level, teachers and students having interest and expertise in evaluation research and IPM will find the book a useful guide for conducting evaluation studies.

For the third time in my research career, I have been closely associated with the dynamic field of IPM and for that, I am highly thankful to Dr. A.K. Dhawan, Professor Entomology (State Coordinator IRM Project, Punjab) at Punjab Agricultural University, Ludhiana, India for providing me an opportunity to conduct the evaluation of the IRM programme in Punjab. The physical and other facilities provided by him are duly acknowledged. His valuable suggestions in the content area of IPM were complimentary to my research. The Senior Research Fellows of the IRM project, Messrs Tripat Kumar, Kamaldeep Singh and Subash Singh were the source of strength, who helped in collecting the data from dawn to dusk, and braved the loo, chilly winds, rains and storms from time to time. The scouts of IRM project, Messrs Avtar Singh, Mewa Singh, Mukhpal Singh and Malkit Singh deserves my thanks. I am greatly thankful to respondent farmers for being cooperative, helpful and kind enough in sparing their valuable time for recording the information in questionnaires, and providing information while administering other research schedules. Remaining with them for the whole cotton growing season was in itself a learning process for me. The guidance and critical analysis of different aspects of research by Dr. Raghbir Singh and Dr. A.J.S. Dhaliwal, Professors of Extension Education (Retd.) inspired and motivated me to work harder. I am thankful to Dr. Dhaliwal for editing the manuscript.

My friends Dr. Kamal Vatta and Dr. Kuldeep Singh deserve special mention for analysis and translation of the data from Punjabi to English. I am thankful to Prof. A.R. Nazki and Prof. A.K. Tiku, SKUAST- Jammu for their help in one form or the other.

I express my regards and gratitude to my parents Mr. M.L. Peshin and Mrs. Uma Peshin, who always showered their love and affection, and I am especially grateful to my father, who inculcated in me the spirit of honesty. Love to my Pummy. My children Udit and Devansh are the fulfillment of my life and future belongs to them–God bless them with all the virtues of life.

Rajinder Peshin

CONTENTS

LIST OF ABBREVIATIONS AND TERMS

ABW	American bollworm (*Helicoverpa armigera*)
Acre	0.4 hectare
ADB	Asian Development Bank
a.i	Active ingredients
Aphids	*Aphis gossypii*
Bt	*Bacillus thuringiensis*
CABI	Commonwealth Agricultural Bureau International
CBA	Cost benefit analysis
CES	Cooperative Extension Service, USA
CICR	Central Institute of Cotton Research, Nagpur, India
CIPMC	Central Integrated Pest Management Centre
CIS	Commonwealth of Independent States
CLCuV	Cotton leaf curl virus
DAP	Di- ammonium phosphate (phosphatic fertilizer)
DAS	Days after sowing
DOF	Department of Finance, Australia
DPPQS	Directorate of Plant Protection, Quarantine and Storage, Government of India

EPA	Environmental Protection Agency, USA
ETL	Economic threshold level
EU	European Union
FAO	Food and Agriculture Organisation of United Nations
FFS	Farmer filed school
g	Gram
ha	Hectare
ha^{-1}	Per hectare
HCH	Hexachlorocyclohexane
ICAR	Indian Council of Agricultural Research, New Delhi, India
ICP	Inter-country Programme of FAO
INRA	National Agricultural Research Institute, Argentina
IPM	Integrated Pest Management
IRM	Insecticide Resistance Management
IRM–farmers	Covered under IRM programme in IRM villages
IRM villages	Where IPM technology was disseminated (experimental group)
Jassid	*Amrasca biguttula*
Kharif	Synonymous with the wet season, however covering the period April/May through September/October, when cotton crop is cultivated in Punjab
KVK	Krishi Vigyan Kendra
m ha	Million hectares
MOP	Muriate of Potash (Potassic fertilizer)
mt	Millions tonnes
non-IRM farmers	Not covered under IRM programme in non-IRM villages
non-IRM villages	Where no IRM programme was conducted (control group)
NRC	National Research Council, USA
NSF	National Science Foundation, USA
NVP	Nuclear polyhydrosis virus
ORP	Operational Research Project
PAU	Punjab Agricultural University, Ludhiana, India
Pest	Includes insect pests, diseases and weeds
Pesticide	Includes insecticides, weedicides, seed treating and agrochemicals

PBW	Pink bollworm (*Pectinophora gossypiella*)
q	Quintals
Rs	Rupees
Sarpanch	Democratically elected village head
SBW	Spotted bollworm (*Earias vittella*)
SD	Standard Deviation
SPSS	Statistical Package for Social Sciences
SSP	Single super phosphate (Phosphatic fertilizer)
TC	Tobacco caterpillar (*Spodoptera litura*)
Thrips	*Thrips tabaci*
UN	United Nations
UNDP	United Nations Development Programme
UNEP	United Nations Environmental Programme
Urea	Nitrogenous fertilizer
USA	United States of America
USDA	United States Department of Agriculture
Wf	Whitefly (*Bemisia tabaci*)

LIST OF FIGURES

LIST OF TABLES

Chapter 1
INSECTICIDE RESISTANCE MANAGEMENT IN COTTON

Cotton is one of the principal commercial crop of India. Though India has the largest area under cotton in the world, it ranks second in production due to low productivity. The percentage of area under cotton is 27 per cent of the world cotton area. The productivity is 300-400 kg lint ha^{-1} against the world average of above 600 kg lint ha^{-1} (Singhal 2003, Singh and Ramasundaram 2005). In India, cotton was grown on about 9 million hectares (m ha) of land up to 2001-02 but the area under cotton reduced around 6.8 m ha in 2002-03 (based on Ministry of Agriculture, Singhal 2003). The strenuous research and development efforts in cotton since independence resulted in remarkable increase in production and productivity through intensive use of chemical fertilizer, irrigation water, adoption of high yielding/hybrid varieties and use of pesticides. The production increased from 3 millions bales in 1950-51 to 13.5 million bales in 2003-04 (Anonymous 2004) and productivity increased from 88 kg ha^{-1} to 238 kg ha^{-1} during the same period (Singhal 2003). The cotton yields increased significantly in the 1980s and through the first half of 1990s, but since 1996 these yields have stagnated and even shown a down ward trend up to 2002-03 (Table 1.1). The low productivity may be due to various factors, but the major factor is the cultivation of cotton crop in rainfed area is most parts of the India.

After the cultivation of *Bt*-cotton since 2002, the productivity has increased from 367 kg ha^{-1} in 2000–01 to 599 kg ha^{-1} in 2007–08 (AICCIP, 2008). However, the productivity of cotton is still lower than the world productivity of 712 kg ha^{-1} (USDA, 2006). One of the major limiting factors for low productivity is the losses caused by insect pests. Central Institute for Cotton Research (CICR 1994, 1998) reported that about 50 per cent of the cotton output in lost due to insect pests, the loss in fact has

Table 1.1: Area and Yield of Cotton in Punjab and India

	Punjab		*India*	
	Area (m ha)	*Yield (q/ha^{-1})*	*Area (m ha)*	*Yield (q/ha^{-1})*
1960-61	0.45	2.69	7.6	1.25
1970-71	0.39	3.71	7.6	1.06
1980-81	0.65	3.09	7.8	1.52
1990-91	0.70	4.63	7.4	2.25
1994-95	0.59	5.02	7.8	2.57
1995-96	0.74	4.47	9.0	2.42
1996-97	0.74	4.41	9.1	2.65
1997-98	0.72	2.20	8.8	2.08
1998-99	0.56	1.79	9.3	2.24
1999-2000	0.48	3.40	8.7	2.25
2000-01	0.48	4.30	8.5	1.91
2001-02	0.60	3.66	8.7	2.08
2002-03	0.45	4.10	7.6	3.38
2003-04	0.45	5.56	7.6	3.09
2004-05	0.51	6.79	8.8	4.70
2005-06[1]	0.56	6.10	8.7	4.72
2006-07[1]	0.59	7.52	9.2	5.20

Sources: Anonymous (2003a, 2005, 2006); Singhal (2003); 1: www.cotcorp.gov.in/statistics.

increased from about 18 per cent in the early 1960s (Pradhan 1964) (Table1.2). The loss of cotton out put to insect pests was valued 31050 million rupees in 1996 (Dhaliwal and Arora 1996) and 137828 million rupees in 2002 (Dhaliwal and Arora 2002). Chandra (1998) and Wahab (1997) estimated that the damage by bollworms was to the tune of 12000 million rupees (about 300 million dollars) every year. Even though use of pesticides in general increased, but still the crop losses after the introduction of pesticides have increased. Under traditional agriculture in India, insect pest losses in cotton crop were estimated to be 18 per cent as per the extensive surveys carried out by various agencies during 1950s (Pradhan 1964) which has increased to around 50 per cent during the 1990s (Dhaliwal and Arora 2001), and based on the 2001-02 prices the losses were estimated to be around 287 600 million rupees (Dhaliwal *et al.*, 2004). This is despite the fact that in India, the average pesticide consumption (technical grade material) increased from 2344 metric tonnes (mt) in 1957-58 (Birthal and Jha 1997) to 75000 mt in 1990-91 (Singhal 2003) to 84095 mt in 1993-94 (Cited from Dhaliwal and Arora 2001). After the ban of HCH and other pesticides and introduction of high potency (lower dosage) new insecticides, abolishing of pesticides subsidies and high priority being accorded to IPM, the consumption of pesticides was estimated to be 54135 mt in 1999-2000 [Directorate of Plant Protection, Quarantine and Storage (DPPQS) Govt. of India 2001]. For the same period per hectare consumption of pesticides (technical grade material) increased from 15.89 g in 1957-58 to 393.69g in 1992-93 (Birthal and Jha 1997) and decreased to 288 in 1999-2000 (Agnihotri 2000). An estimated 44.5 per cent (Table 1.3) of this is being consumed in cotton (DPPQS 2007a)[1]. The pesticide consumption (insecticides, fungicides, nematicides and herbicides) have shown the downward slide since 1991-92 (Table 1.4) and this is due to elimination of pesticide subsidies, development of newer molecules which require less dosage per unit area and also due to implementation of IPM programmes whose coverage has reached five per cent of total cropped area.

Table 1.2: Crop Losses Caused by Insect Pests in Cotton Crop in India

Year	*Loss %*	*Value (Million Rs)*
1964[1]	18.0	NA
1986[2]	7.0	500
1996[3]	22.0	20000
1996[4]	50.0	31050
2002[5]	50.0	137828
2004[6]	50-60	287 600

Sources: 1: Pradhan(1964); 2: Atwal(1986); 3: Lal(1996); 4: Dhaliwal and Arora (1996); 5: Dhaliwal and Arora (2002); 6: Dhaliwal *et al.* (2004).

1.1 Integrated Pest Management (IPM) in Cotton

The main reliance on the use of insecticides led to creation of newer pest problem in all the crops and especially in cotton crop. Due to lack of resistant cultivars, non-adoption of cultural control measures, and non-availability of effective bio-agents, the indiscriminate use of insecticides resulted in development of resistance in cotton pests like American bollworm (ABW), resurgence of pests like whitefly and destruction of natural enemies which many a times resulted in failure of cotton crop. Such type of cotton failure has been documented in Latin America (Cänete Valley Peru), Sudan and other places even before the term IPM was coined.

Integrated Pest Management is one of the components of sustainable agriculture along with integrated nutrient management, soil conservation, and water management system (Pretty 1995). IPM has been defined by the Panel of Experts on Integrated Pest Control at Food and Agricultural Organisation (FAO), Rome, as:

"A pest management system that, in the context of the associated environment and the population dynamics of the pest species, utilizes all suitable techniques and methods in a compatible manner as possible and maintains the pest population at levels below those causing economic injury" (FAO 1968).

---

[1] Directorate of Plant Protection, Quarantine and Storage (DPPQS) Govt. of India (2007a).

IPM is integration of different (available) pest management practices in a compatible manner, so as to keep pest population below economic injury level, in such a way that it is not only economically viable but also ecologically sound. The word 'available' has been added to denote different pest management tactics available in a particular area and under a particular IPM system. The objective of IPM is to reduce the pesticides use and to maintain ecological balance there by generating sustainable agricultural growth.

Table 1.3: Crop-wise Percentage Share of Area and Consumption of Pesticides in India

Sl.No.	*Crop*	*% of Total Pesticides Consumption**	*% Share of Cropped Area** (1991-92)*
1.	Cotton	44.5	4.20
2.	Paddy	22.8	23.30
3.	Jowar	8.9	6.80
4.	Wheat	6.4	12.80
5.	Tuar	2.8	2.00
6.	Groundnut	2.5	4.70
7.	Bajra	1.8	5.60
8.	Maize	1.1	3.20
9.	Sugarcane	0.7	2.20
10.	Ragi	0.4	1.20
11.	Jute	0.3	0.50
12.	Gram	0.2	3.10
13.	Onion	0.2	0.20
14.	Rapeseed and Mustard	0.2	3.50
15.	Tobacco	0.1	0.30
16.	Sunflower	0.1	1.16
17.	Potato	0.1	0.60
18.	Soybean		1.74
19.	Tapioca	6.9	0.14
20.	Others@		22.76
	Total	100.00	100.00

*: Dudani and Sengupta, 1991

**: Agriculture Statistics at a Glance, DAC (March, 1995)

@: Include all fruits and vegetables etc. which are not included in the above 19 crops.

Different projects/programmes related to IPM were launched in India from time to time to reduce pesticides consumption, its adverse effects, and to reduce losses caused by cotton pests. The IPM programmes implemented in India are:

Operational Research Project in 1975; Regional Programme on cotton-IPM by Commonwealth Agricultural Bureau International (CABI) in 1993; FAO-European Union IPM programme for cotton in 2000; National Agricultural Technology Project (NATP) for IPM in 2000. There are many organisations involved in implementing different IPM programmes in India. The CICR, Nagpur also launched IPM programme in 10 cotton growing states of India called Insecticide Resistance Management (IRM) in 2002.

Table 1.4: Consumption of Pesticides for Last 21 Years

Sl.No.	*Year*	*Consumption M.T. (Tech. Grade)*	*Consumption of Pesticides per ha (Grams)*
1.	1985-86	61881	–
2.	1986-87	67270	–
3.	1987-88	66895	–
4.	1988-89	75418	–
5.	1989-90	71894	–
6.	1990-91	75033	404
7.	1991-92	72133	392
8.	1992-93	70794	381
9.	1993-94	63651	341
10.	1994-95	61357	326
11.	1995-96	61260	327
12.	1996-97	56114	296
13.	1997-98	52239	274
14.	1998-99	49157	255
15.	1999-2000	46195	243
16.	2000-01	43584	229
17.	2001-02	47020	247
18.	2002-03	48350	254
19.	2003-04	41020	–
20.	2004-05	40672	–
21.	2005-06	39773	–

Source: Directorate of Plant Protection Quarantine and Storage, Government of India (2007a).

1.2 Insecticide Resistance Management (IRM): A Component of IPM

The focus of IRM programme is on rationalising insecticides use. IRM approach avers that the contribution of insecticide resistance to pest control failures has been under estimated. The fact is that unless full-fledged efforts are made to understand all aspects of the resistance phenomenon, any attempt to implement IPM at field level would be peripheral (Bambawale *et al.*, 2004). IRM is a strategy of using few application of any insecticide as possible, in order to delay the development of resistance in insects. The strategy includes spraying only when economic threshold levels (ETL) are researched by using most selective insecticides to start with, so as natural enemies will survive and control the pest population, and rotating between different insecticides classes. This strategy is presented within a full IPM context (Russell 2004). The promotion of other IPM practices, such as adoption of sucking pest tolerant varieties, seed treatment, use of all available tactics for management of a particular insect pest like timely sowing, clean cultivation, etc are the components of this programme.

In the absence of effective bio-agents and alternative chemistries, and the development of resistance in insect pests particularly American bollworm *Helicoverpa armigera*, the tactics for overcoming this insecticides resistance were evolved. Results were so positive that by 1999, the programme was being implemented in four states of India. The IRM programme was started during IX-Plan Period under Technology Mission on Cotton which was extended up to 2007, covering 28 districts, in ten states, which, between them consume 80 per cent of India's insecticides use on cotton (Russell 2004).

The module for IRM based IPM strategy has been developed by CICR, Nagpur. The objectives of the IRM project are:

1. To disseminate available technology related to IRM,
2. To monitor the level of resistance to commonly used insecticides,
3. To assess the impact of IRM strategy on economic status of the farmers, and
4. To validate the available IRM strategy.

The technology disseminated under IRM programme is based on IPM principles. The technology designed for dissemination includes: Adoption of general agronomic practices with the aim to reduce the use of insecticides and managing insecticide resistance, farmers' training in the identification of insect pests and their natural enemies, economic threshold level (ETL) application in spray decisions, and to create awareness among farmers about development of insecticides resistance.

1.2.1 Cultural Practices

1. Growing of recommended early maturing varieties of cotton.
2. Destroying alternate hosts of insect-pests *viz. Kanghi buti, Peeli buti, Itsit*.
3. Seed treatment.
4. Timely (April) sowing of cotton crop
5. Recommended fertilizer dosages and use of farmyard manure (FYM).
6. Seed treatment and seed rate
7. Management of recommended plant spacing
8. Judicious and timely application of irrigation water

1.2.2 Chemical Practices Based on IRM Strategy

Days After Sowing (DAS)		*Action*
0-60	—	No insecticide use. In case of emergency, if needed, use endosulfan 35EC, effective against jassid also.
60-90 (1st window)	—	Use of endosulfan 35EC. In case of no effective control by endosulfan 35EC, use synthetic pyrethroids for spotted bollworm and American bollworm (early instars) in case of emergency, no organophosphates/carbamates/pyrethroids until 90 DAS.
	—	Nitroguadine compounds can be used if endosulfan 35EC fails to control jassid.
90-110 (2nd window)	—	Use of endosulfan 35EC not beyond 90 DAS uses synthetic pyrethroids/organophosphates/carbamates against spotted bollworm. Use profenophos/quinalphos/triazophos for young larvae; chlorphyriphos/acephate for grown up larvae of American bollworm.
110-140 (3rd window)	—	Use of profenophos/quinalphos/triazophos for young larvae, chlorphyriphos/acephate for older larvae of American bollworm or spinosad/indoxacarb if above insecticide fail to control grown up larvae of American bollworm.
	—	Use triazophos or ethion for control of whitefly.
	—	Use chlorphyriphos/acephate/endosulfan or quinalphos for control of tobacco caterpillar.
After 140 (4th window)	—	Use of chlorphyriphos/indoxacarb/spinosad/quinalphos as emerging operation against American bollworm
	—	Use ethion/triazophos for control of whitefly.

The pesticide application should be based on economic threshold levels (ETL). The ETL for different insect pests are: 5 per cent damage in shed fruiting bodies for bollworms, appearance of yellowing and curling along leaf margins on 50 per cent plants in case of jassid, and 6 adults per leaf or appearance of honeydew on 50 per cent plants for whitefly.

1.2.3 Conservation of Natural Enemies

Conservation of natural enemies by not using any pesticides up to 60 days after sowing, and to avoid broad-spectrum pesticides until 90 days after sowing.

1.3 Cotton Growing Scenario in Punjab

The area under cotton in Punjab was estimated to be 0.45 m ha during 2002-03 while it fluctuated between 0.397 m ha to 0.701 m ha since 1960-61 to 2002-03

(Table 1.1). During the same period, the yield of American cotton varied between 179 kg ha^{-1} to 481 kg ha^{-1} (Anonymous 2003a). The cotton crop in Punjab was adversely hit by bollworms in 1998 and 2001 and this was this was attributed to long wet spell resulting in high vegetation, and excessive use of synthetic pyrethroids and nitrogenous fertilizers (Singhal 2003). Synthetic pyrethroids were responsible for elimination of natural enemies. The overall yield reduction took place in northern zone cotton growing states of Punjab, Haryana and Rajasthan. Pesticide consumption in Punjab was 5770 metric tones in 1988-89, which increased to 7600 metric tones in 1995-96, and was estimated to be 7100 metric tons in 1999-2000 (Dudani and Sengupta 1991, Puri 1995, Directorate of Plant Protection Quarantine and Storage, 2001) (Figure 1.1). The overall consumption of pesticides decreased to 5710 metric tons in 2006-07(Table 1.5) (Directorate of Plant Protection Quarantine and Storage, 2007b)[2]. The consumption of insecticides (technical grade material) in Punjab is 923g ha^{-1} (Agnihotri 2000), which is the highest in India. Around fifty per cent of the total pesticide consumption is on cotton crop.

The cost of insecticides to total cost of cotton production has risen alarmingly. In 1974-75 it was 2.1 per cent of cost of cultivation, in 1979-80 it rose to 4.6 per cent, in 1984-85 it was 11.9 per cent, in 1989-90 it rose to 15.5 per cent and in 1994-95 it declined to 13.0 per cent (Dhaliwal and Arora 2001). In 1998-99 it was estimated to be 21.21 per cent (Figure 1.2) (Sen and Bhatia 2004). Shetty (2004) in a study on pesticide hot spots estimated it to be 50 per cent in case of district Bathinda in Punjab.

This entire scenario presents a grim picture of exploitation of natural resources by farmers, and farmers being the ultimate sufferers of this. A sizable number of farmers had given up the cotton cultivation after 1995-96 till 2004 in some of the villages due to ravages caused by insect pests mainly American bollworm to the cotton crop. The insect pest scenario under went a change during the last three decades. In 1970s, there were only two major pests of cotton crop, namely jassid and pink bollworm. But more than six major pests of cotton were reported in 1990s, namely jassid, whitefly, spotted bollworm, American bollworm, pink bollworm and cotton leaf curl virus (Table 1.6). IRM programme was started in Punjab during 2002 and 30 villages were covered under this programme. During 2003 and 2004, 30 and 45 villages, respectively were adopted under this programme. Some of the villages covered during 2003, were repeated in 2004. During 2005, a total of 120 villages have been adopted under IRM programme and most of the villages covered earlier have also been included during 2005.

1.4 Evaluation of IRM Programme

IPM implementation in India is mostly based on instructing the farmers rather than educating them (Ooi *et al.*, 1992, FAO 1996). IPM being complex, interacting technclogy, can be successfully implemented at farmers' level by adopting

---

[2] Directorate of Plant Protection Quarantine and Storage, Government of India (2007b).

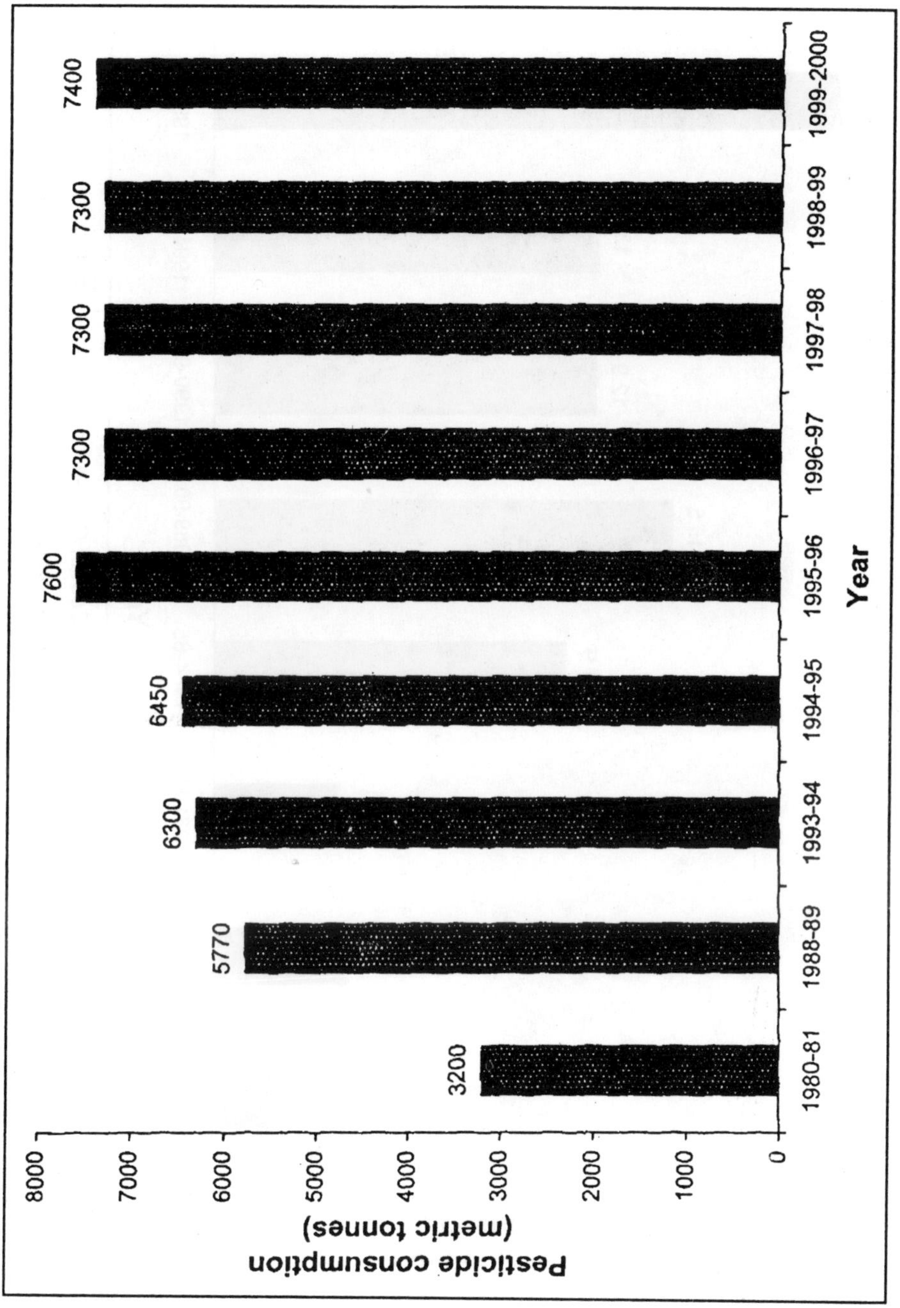

Figure 1.1: Pesticide Consumption (Tech. Grade Material) in Punjab
***Sources*: Dudani and Sengupta (1991), Puri (1995) and DPPQS (2001)**

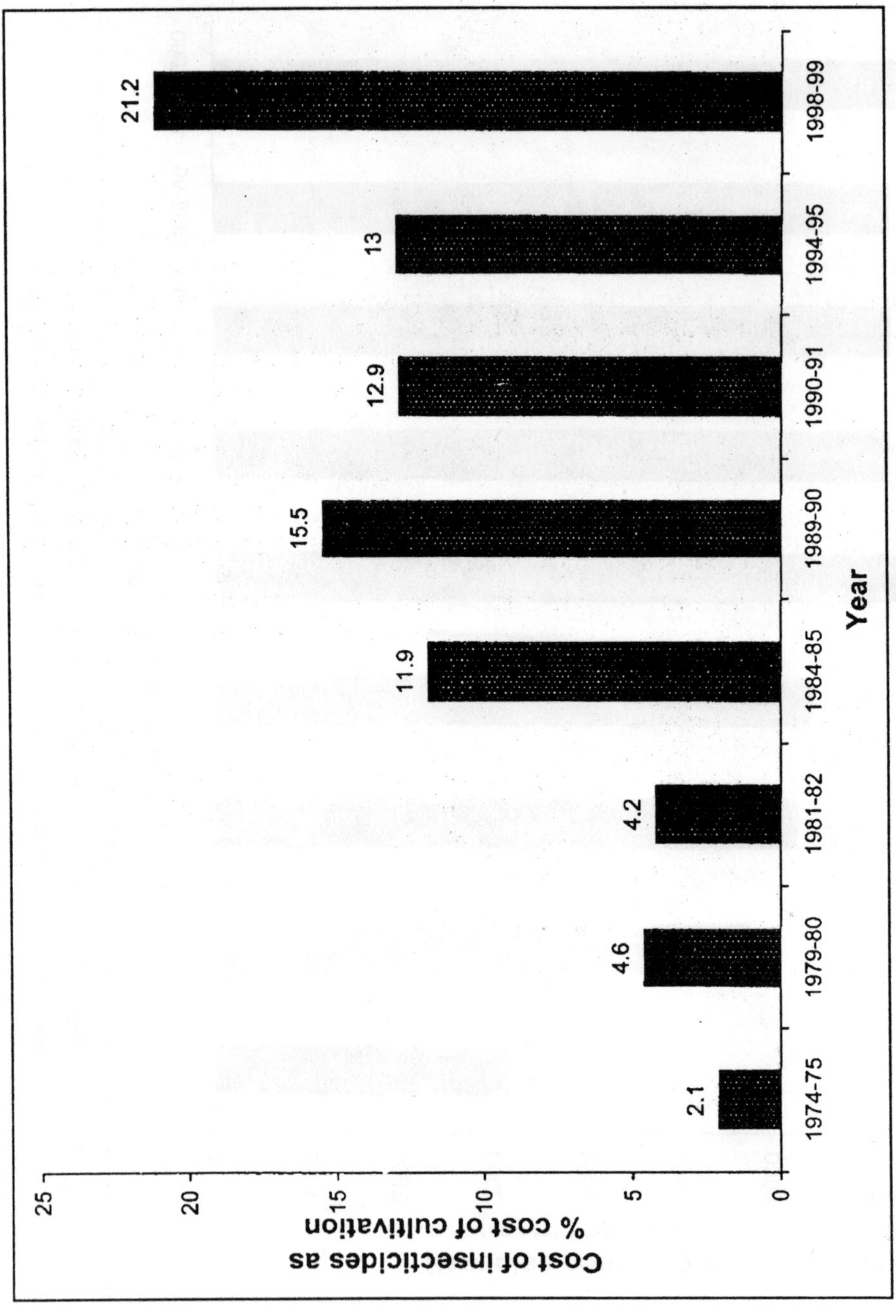

Figure 1.2: Cost of Insecticides as Per cent Cost of Cotton Cultivation in Punjab

***Sources*: Dhaliwal and Arora (2001), Sen and Bhatia (2004)**

Table 1.5: Consumption of Pesticides in Various States During 2002-07

M.T. (Tech. Grade)

Sl.No.	*States*	*2002-03*	*2003-04*	*2004-05*	*2005-06*	*2006-07*
1.	Andhra Pradesh	3706	2034	2133	1997	1051
2.	Assam	181	175	170	165	165
3.	Arunachal Pradesh	15	147	17	02	17.10
4.	Bihar	1010	860	850	875	890
5.	Chhattisgarh	-	332	486	450	490
6.	Gujarat	4500	4000	2900	2700	2670
7.	Goa	05	05	05	05	6.90
8.	Haryana	5012	4730	4520	4560	4600
9.	Himachal Pradesh	380	360	310	300	290
10.	Jammu and Kashmir	98	09	12	1433	829
11.	Jharkhand	40	56	69	70	57
12.	Karnataka	2700	1692	2200	1638	620
13.	Kerala	902	326	360	571	547
14.	Madhya Pradesh	1026	662	749	787	879
15.	Maharashtra	3724	3385	3030	3198	3193
16.	Manipur	19	25	26	28	27
17.	Meghalaya	06	06	08	06	9.30
18.	Mizoram	15	15	25	25	40
19.	Nagaland	07	07	05	05	05
20.	Orissa	1134	682	692	963	1068
21.	Punjab	7200	6780	6900	5610	5710
22.	Rajasthan	3200	2303	1628	1008	1523.20
23.	Sikkim	03	03	No as Organic State	No as Organic State	02
24.	Tamil Nnadu	3346	1434	2466	2211	2048
25.	Tripura	88	118	17	14	09
26.	Uttar Pradesh	6775	6710	6855	6672	7022
27.	Uttarakhand	129	147	132	141	139
28.	West Bengal	3000	3900	4000	4250	3950
29.	A and N Islands	03	06	03	03	4.415
30.	Chandigarh	01	0.78	0.78	0.78	N/A
31.	Delhi	60	56	53	39	57
32.	Dadra and Nagar Haveli	05	05	05	04	N/A
33.	Daman and Diu	01	01	01	01	N/A
34.	Lakshadweep	02	02	02	01	N/A
35.	Pondicherry	57	46	42	41	4.46
	Total	48350	41020	40672	39773	37959.375

Source: Directorate of Plant Protection Quarantine &Storage Government of India (2007b) based on Zonal conference on Inputs, *Kharif*, 2007.

participatory extension methodology to educate farmers. Formative evaluation of the IRM programme was planed to evaluate the project objectives related to dissemination of IRM technology, to assess the economic benefits of the programme at farmers' level and to provide feed back to the stakeholders of programme for its improvement. As such the study entitled "Evaluation of Insecticide Resistance Management Technology in Cotton Crop in Punjab" was undertaken.

Table 1.6: Changing Pest Scenario on Cotton in North Zone of India

Year	*Major Pest*	*Minor Pest*	*New Pest*	*Pest Outbreak*
Upto 1970	J, PBW	Wf, Ap, Th	SBW	
1971-1980	J, PBW, SBW	Wf, Ap, Th	ABW	ABW in some pockets (1978)
1981-1985	J, PBW, SBW	Wf, Ap, Th	-	ABW (1983)
1986-1990	J, SBW, PBW	Wf, Ap, Th	TC	ABW (1990)
1991-1995	J, SBW, ABW, PWB	Wf, Ap, Th	LM, CLCuV	ABW (1995); WF (1995)
1996-2000	J, Wf, SBW, ABW, PBW, CLCuV	Ap, Th	-	ABW (1997, 1998)
2001-2003	J, Wf, SBW, ABW	Ap, Th, Tc	-	ABW (2001) TC in some pockets (2003)

J: Jassid; Wf: Whitefly; Ap: Aphids; Th: Thrips; PBW: Pink bollworm; SBW: Spotted bollworm; ABW: American bollworms; Tc: Tobacoo caterpiller; LM: Serpentine leaf miner; CLCuV: Cotton leaf curl virus.

Source: Dhawan *et al.* (2004).

1.5 Objectives of the Study

1. To evaluate the farmers' training under IRM project in terms of training strategy and knowledge gain of farmers.
2. To find out the adoption, adoptability and economic benefits of IPM practices in cotton crop disseminated under the IRM project.
3. To find out the constraints in the adoption of IPM practices disseminated under the IRM project.

1.6 Significance of the Study

As already stated, this study is formative/process evaluation of the IRM programme and aims to provide feed back to multiple stakeholder groups of the programme. The evaluation was conducted during the third year of programme implementation phase. It will provide the feedback, whether the programme is implemented as planned and whether inputs result in the expected output through the adopted process of implementation.

The study analyses the knowledge gain of farmers about IPM by employing pre and post-training knowledge test and such information will provide the necessary feed back about the different aspects where knowledge gain has taken place and point out those aspects of IPM in which desired knowledge gain has not been achieved, and which needs to be looked into by the trainers. The information generated about the adoption of different IPM practices, and the comparisons made with non-IRM villages will serve as the guidelines for future implementation of the programme.

The study also analyses the economic benefits occurring to the participating farmers in IRM programme and the reduction and judicious use of insecticides. This shall help the stakeholders of the programme to analyse whether desired output is achieved and if not what steps and methodology needs to be followed for making training more meaningful. The IPM practices being disseminated under IRM programme have to be analysed in terms of their adoptability at farmers' level. This will provide the necessary inputs to the biological scientists as to which practices or innovations farmers are likely to adopt and reorient their strategy accordingly.

Evaluation of the IRM programmes by trainee farmers will give the most important insight into the short-comings of the programme, and what needs to be done to remove these. The study has also highlighted the constraints faced by the farmers in adopting IPM, and at what level IPM integration has taken place.

Quintessentially the study will be useful to stakeholders of the programme, research scientists, extension agencies and ultimately to the end users of IPM.

1.7 Limitations of the Study

One of the limitations of this study is that the experimental research design employed for the study was faulty, as in case of control group no pre and post knowledge test was administered. The other limitations of the study are:

1.7.1. The formative evaluation of the programme might have influenced the trainers and quality of training in the study villages, as such generalization for all IRM villages can not be made without considering this aspect.

1.7.2. The adoptability of the selected IPM practices was measured by developing a scale and working out the adoptability indices but the adaptability (active adoption) could not be quantified to predict the adaptability of selected IPM practices.

1.7.3. The base line data of 2003 was collected by personnel interview method; there can be some error on account of farmers' ability to recall/recollect the practices adopted by them in terms of pesticide use and expenditure. There might be some discrepancies between the actual information and expressed response.

1.7.4. The number of tank mixtures of pesticide applications applied in 2003 (base line data) were not known as most of the farmers were not able to recollect the number of tank mixture used in 2003. Therefore, before IRM and during IRM data were compared by accounting tank mixture in 2004 as one application, which has given conservative estimates about the reduction in insecticide applications.

1.7.5. The overall sample size of the IRM villages selected for the study was small, as out of 45 villages covered under the programme in 2004 only 15 villages were taken up for the study.

1.7.6. The study could not be continued to assess terminal impact (after IRM) of the IRM programme due to shortage of time and resources as is the case with most of the students' research programmes.

Chapter 2

INTEGRATED PEST MANAGEMENT

Implementation and Impact

The main purpose of literature review is to set the research study within a wider context to show how the study supplements the work that has already been done on IPM, and its dissemination and adoption. The review of literature helps in formulating the strategy and data collection methods, if working deductively. On the other hand, if working from data to theory (inductively) literature review serves the purpose of illuminating and enriching the conclusions (Saunders *et al.*, 2004). The theoretical rationale of the process area has been discussed in chapter 3. In this chapter theoretical rationale of the IPM, its development, dissemination, adoption, economic benefits and other related aspects are discussed with respect to the existing body of research on the subject. Therefore, an attempt has been made to review the relevant literature under the following headings:

2.1 IPM: A historical perspective

2.2 Development of IPM

2.3 IPM programmes around the world

2.4 Knowledge achievement: A learning process

2.5 Adoption of IPM practices

2.6 Economic rewards of IPM programmes

2.7 Adoptability and constraints

2.1 IPM: A Historical Perspective

The basic tactics of IPM were proposed and applied to reduce crop losses against the ravages of pests long before the expression was coined (Jones 1973, Smith *et al.*, 1973). Throughout the early twentieth century, plant protection specialists relied on knowledge of pest biology and cultural practices to produce multitactical control strategies (Gaines 1957). Pest control was understood as the set of actions taken to avoid, attenuate, or delay the impact of pests on crops, as such goals and procedures of pest control were clearly understood (Kogan 1998). With the introduction of synthetic pesticides in 1940s, the whole scenario changed. The over reliance on synthetic pesticides from late 1940s to mid 1960s was the era of optimism. The period was dominated by the belief that pesticides were the solution to pest management. In 1960s itself, a major change in the thinking occurred following the realization that pesticides had serious limitations and in several cases produce undesirable consequences. Resistance of pests to pesticides was observed during this period, phenomenon of pest resurgence and development of minor pests to major pests due to killing beneficial insects was documented in late twentieth century (Norris *et al.*, 2002). Rachel Carson (1962) wrote a book *Silent Spring*, that brought the problems caused by pesticides to the attention of the public and forever changed the way pesticides are viewed and dealt with by scientists. Stern *et al.* (1959) coined the concept of integrated control and the publication of '*Silent Spring*' accelerated the concept of integrated control. However, not until 1972, were "integrated pest management" and its acronym IPM incorporated into the English literature and accepted by the scientific community (Kogan 1998) and later, in November 1972, the report *Integrated Pest Management* prepared by the Council on Environmental Quality was published (Anonymous 1972).

2.2 Development of IPM

The term IPM is now more or less universally understood. Even before the term IPM was coined, the reasons for developing and propagating IPM are explained by citing some well documented historical cases.

Cänete Valley, Peru had been a successful cotton growing area with progressive farmers. In 1939, the tobacco bud worm (*Heliothis virescens*) appeared in cotton crop. The spraying of arsenical insecticide and nicotine sulphate resulted in build-up of cotton aphid (*Aphis gossypii*) and worsening of tobacco bud worm. By 1949, cotton yields (lint) dropped from about 500 kg ha^{-1} to 365 kg ha^{-1} as natural enemies had disappeared owing to insecticide applications. A new programme for pest control practices was introduced which included banning of synthetic organic pesticide use, the reintroduction of beneficial insects, crop diversification schemes, planting of early maturing varieties and the destruction of cotton crop residues. Pest problem declined dramatically and pest control costs were substantially reduced (Hasen 1987).

Based on the same principles as IPM, efforts were for 'harmonious control' in Canada in 1950s (Pickett and Patterson 1953; Pickett *et al.*, 1958). The concept of integrated control in the USA was developed in late 1950s and it consisted mainly of the use of insecticides in a manner that was compatible with biological control of insect pests (Norris *et al.*, 2002). The cotton growing in Sudan also suffered due to over reliance on insecticides. DDT induced outbreaks of cotton whitefly and brought about occurrence of American bollworm (*Heliothis armigera*) which resulted in reduction in yields (Joyce and Roberts 1959).

In Nicaragua, after a successful production phase in 1960s, the average yields of cotton crop fell by 15-30 per cent, despite 28 insecticide applications per season. In 1971, a programme was started which disseminated practices like 'trap cropping' and use of selective insecticides on weevil trapped in trap crop (Swezey *et al.*, 1986).

All these developments in crop protection have been driven by changing pest problems faced by the farmers, the options available to them and their changing cash and labour requirements (Norton 1993). Thus the developments of IPM and with it started a search for a perfect definition. A broader definition was adopted by the FAO Panel of Experts in 1968 (refer Chapter 1). A survey has recorded 64 definitions of IPM and the key words included in those 64 definitions suggests that authors attempted to capture (a) the appropriate selection of pest control methods, used singly or in combination; (b) economic benefits to growers and society; (c) the benefits to the environment; (d) the decision rules that guide the selection of the control action, and (e) the need to consider impact of multiple pests (Kogan 1998).

The theory and principles supporting IPM have been developed over the last 50 years. Integrated control, the term first applied to IPM, was developed and introduced as a concept in the USA in the late 1950s. It was developed to harmonize chemical and biological control (Smith and Allen 1954). This early concept was based on premise that pesticides could have a minimum impact on the natural enemies of the pest if applied at correct time and under the correct conditions. Economic threshold, another important concept in IPM, was introduced at that time (Anonymous 1999). These concepts remained the major themes of IPM throughout much of the 1970s. The United Nations Development Programme (UNDP) together with the FAO since 1975 initiated global programmes for the development and application of IPM programmes in rice, cotton, sorghum and millet (Oudejans 1991). The focus of IPM began to shift to non-pesticidal tactics in the 1980s, including expanded use of cultural control, introduction of resistant varieties and biological control.

2.2.1 Development of IPM in India

In India, the research on integrated pest management was started in 1974-75 on two crops, rice and cotton under Operational Research Projects (Swaminathan 1975). Under this location specific IPM technologies were developed in cotton and rice crops. The Punjab Agricultural University, Ludhiana also started giving 'integrated pest control schedule' for cotton crop, in its package of practices in late 1970s. The economic threshold level of jassid in cotton crop was included in package of practices by PAU in 1979. But it was only in mid 1980s that Government of India reoriented its plant protection strategy. India became a member country of FAO initiated Inter-Country Programme in 1980, but IPM activities were intensified only since 1993.

2.3 IPM Programmes Around the World

Several events in early 1970s converged to help advance the IPM cause in the United States of America (USA) and other countries. The spread of IPM programmes in the world mainly began in 1980s.

2.3.1 IPM Programmes in the USA

'Federal IPM Trust' was created in the USA in 1972. The United States Department of Agriculture (USDA), National Science Foundation (NSF) and Environmental Protection Agency (EPA) jointly financed 5- years programme of IPM to cover around 1.6 million hectares (Kogan 1998), known as the Huffakar Project. Six crops viz–alfalfa, citrus, cotton, pines, pome and stone fruits and soybean were covered under the project (Huffakar and Smith 1972). Project spanned 1972-78. The second large scale project was launched, known as Consortium for Integrated Pest Management (Frisbie and Adkisson 1985). Regional IPM programme was launched with the consortium for IPM which concluded in 1985. National IPM initiative for implementing IPM practices on 75 per cent of the USA's crop area by 2000 was started in 1993 (Sorensen 1994). The American Cooperative Extension Service (CES) plays a key role in dissemination of IPM in the United States (Frisbie 1994). The IPM programmes evolved expanded to include entire crop pest complex, and there is greater emphasis on multidisciplinary team approach to IPM, with CES and research cooperating at all phases of programme development, implementation, and evolution (Kogan 1998).

2.3.2 IPM Programmes in Latin America

During the late 1970s, the agricultural research and extension services in Brazil initiated an intensive programme to transfer IPM technology to cotton farmers (Bleicher *et al.*, 1979). Later, chemical pesticide industry began to collaborate in IPM programme which did not show desired results (Ramalho 1994). In Peru, the Peruvian Action Network of Alternatives to Agrochemicals (RAAA) in 1992 started a training programme in Cänete Valley to reawaken farmers' interest in IPM, under the theme 'Ecological Pest Management for Cotton Growers'. In 1997, a small project on organic cotton production was also set up (CABI 2000). There is no government extension service in Peru. In Colombia, a growers' cooperative with 180 members in 1997 started working on 600 hectares to start IPM programme and in 1998 it spread to 2400 hectares (Williamson 1999). The National Agricultural Research Institute's (INRA) cotton IPM programme in Argentina is researching and implementing, mass production of predator *Chrysopa* spp. for aphid and cotton pest management. The IPM methods have succeeded in reducing insecticide application from 11-12 per season to less than 4 (Williamson 1999). The only attempt to organise farmer training along discovery–learning methods are in Peru. Most training in IPM programmes in Latin America is based on result demonstration methods with little active farmer participation.

2.3.3 IPM Programme in Africa

Egypt in the early 1970s developed an effective integrated pest control recommendations consisting of cultural, biological, chemical and regulatory measures.

These practices, is combination with some others have been very successful. Average number of sprays in 1975 and later years was less than one. In Sudan after the whitefly problem in 1979, when the problem spread out of control, a programme for development and application of integrated pests control in cotton was implemented by FAO and financed by the Government of Netherlands. The first phase was 1979–83 and second phase started in 1985 to 1989 (Oudejans 1991).

2.3.4 IPM Programme in Commonwealth of Independent States (CIS)

Major areas of cotton production in CIS are Uzbekistan, Turkmenistan, Azerbaijan and Tajakistan (Sugonyaev 1994). The IPM programmes for cotton in the CIS have been developed on the basis of their practicality and economic expedience. IPM programmes are flexible, open system aimed at achieving ecological stabilization. Natural enemies are considered component of IPM programmes as these suppress 60-70 per cent of pest population (Niyazov 1992). The pesticides shortages and dramatically increased costs (unlike in former USSR), coupled with public concern have created an environment for rapid progress of IPM.

2.3.5 Australian IPM Programmes

Professional cotton consultants, mostly direct pest management in Australian cotton. Large corporate farms employ their own trained staff, 90 per cent of individual growers employ a consultant who check crops and advises on pest management (Fitt 1994). Until early 1990s, SIRATAC Ltd. provided unique umbrella for the dissemination of research results through a central computer at Narrobi Agricultural Research Station (Brook and Hearn 1990). The agrochemical industry played major role in the provision of services to growers and in the implementation of IRM. The emphasis shifted from IRM (which was mainly based on reliance on chemical control) to sustainable and effective IPM. IPM practitioners weaned away growers following pesticides industry controlled IRM strategy and the percentage of growers adopting IPM had grown from zero to 30 per cent in 2000 (CABI 2000). The grower participatory IPM has increased from the development of the IPM guidelines to the formation of IPM groups (Wilson *et al.*, 2004).

2.3.6 IPM Programmes in Asia

The role of FAO in dissemination of IPM is well documented. FAO provided the coordination, leadership and resources to promote IPM, particularly in developing countries. From 1977 to 1987 IPM moved from research towards extension. By 1988, Training and Visit extension system in Philippines, Indonesia, Sri Lanka, Bangladesh, India, Thailand and Malaysia attempted to introduce IPM to rice farmers through their system of 'impact points' or through strategic extension campaign (Kenmore 1997). From 1988 to the present IPM, has moved towards education rather than training. 'A Global IPM Facility' with co-sponsorship of FAO, UN Development Programme (UNDP), UN Environmental Programme (UNEP), and the World Bank was established (Kogan 1998). 'Global IPM Field Exchange and Meeting' was held in 1993, where participants from Africa, the near East, Latin America, and Europe observed the success of success of Asian IPM farmers in South-east Asia (Kenmore 1997). The FAO–European Union IPM programme, for cotton in Asia was established in late 1999. The programme worked in six countries: Bangladesh, China, India, Pakistan, Philippines and Vietnam (Ooi 2003).

In India, CICR, Nagpur; Asian Development Bank (ADB)–Commonwealth Agricultural Bureau International (CABI) and Directorate of Plant Protection Quarantine and Storage, Government of India conducted season–long trainings for IPM–extension workers since 1994 to promote IPM (Bambawale *et al.*, 2004). Central Integrated Pest Management Centers (CIPMCs) were set up in 26 states which promoted the concept of IPM in cotton and rice since 1990s. Various state departments of agriculture implemented IPM from mid–nineties. From 2000–2001, Government of India launched the Technology Mission on Cotton (Barik *et al.*, 2002). FAO–EU launched IPM programme in cotton in India since 2000 for five years. Andhra Pradesh cotton IPM initiative is another active organisation in IPM (Anonymous 2001). Multilocation trials have been carried out (including Punjab) by All India Coordinated Cotton Improvement Project (Anonymous 2004). The *Ashta* IPM model is also being implemented in Central India. Agriculture Man Ecology (AME) funded by a bi-lateral agreement between India and Dutch governments is implementing IPM farmer field schools in Karnataka, Andhra Pradesh and Tamil Nadu. IRM is another IPM programme being implemented all over India. IRM approach is based on the premise that unless full-fledged efforts to understand all aspects of resistance phenomenon, any attempt to implement IPM at field level would not bear results. Networked research from 1993, between Indian institutions and the UK, resulted in understanding the mechanisms of resistance (Kranthi *et al.*, 2001a), the extent of problem and of the seasonal variations in the efficacy of particular product (Kranthi *et al.*, 2001b). Positive results led to Government of India in expanding the programme from four states to ten major cotton growing states, focusing on the 26 districts, which between them, are consuming 80 per cent of insecticide use in cotton (Russell 2004). Although, as discussed in chapter1, the main focus of the programme is on rationalising insecticide use, this is presented within a full IPM context.

In India, many agencies are involved with the implementation/dissemination of IPM technology, but the area covered under IPM is less than five per cent (Ragunathan 2005), and without any empirical impact evaluation of these programmes in India. The actual spread of IPM practices being adopted by farmers is not well documented as was also pointed out by Luttrell *et al.* (1994) in a comprehensive review of cotton IPM systems of the world. The success of different IPM programmes depends upon the widespread adoption of IPM practices by the farmers and for that learning for empowerment of farmers is prerequisite. Otherwise these programmes shall remain confined to projects and project reports.

2.4 Knowledge Achievement: A Learning Process

Knowledge gain of farmers about IPM practices is considered the first reward of IPM programmes (Rossing *et al.*, 1985; Zadoks 1989a, van de Fliert 1993, Peshin and Kalra 2000a). The phases for impact assessment of IPM programme are natural, human, social and financial/physical. Under the human phase (farmers) the indicators are participation, IPM knowledge, IPM skills and IPM practices. Participation of farmers in these programmes is a basic indicator which would result in gain of knowledge and skills about IPM practices. Mostly the studies have been confined to study cognitive domain and that too only up to knowledge level. Most of such studies indicated that farmers participating in IPM programmes have gained in knowledge.

Kenmore *et al.* (1987), Medina (1987) and Pincus (1991) studied IPM training programmes and indicated that farmers acquired knowledge through these programmes. Adalla and Rola (1986) found that on farm experimentation involving farm participation in Philippines resulted in farmers' acquiring knowledge through IPM training programme. Haque *et al.* (1988) administered a performance test following the training course. The performance test showed significant increase in the knowledge of participants about IPM principles. Oudejans (1991) explained that small scale farmers had little knowledge either about disease problems, or about selecting the right product, dosage and timing for collecting individual pests. Bhuiyan *et al.* (1992) in their study on the use of ETL found that higher percentage of IPM trained farmers (100 per cent) compared to untrained farmers (80 per cent) were aware of the natural enemies of pests. Rola and Baril (1997) evaluated the impact of FFS in developing knowledge based technologies in Philippines. The FFS trained farmers had knowledge of the concept and practical application of the agro ecosystem analysis.

van de Fliert (1993) studied the IPM-farmer field school (IPM-FFS) programme in Indonesia and used pre and post-ballot box test to study the gain in knowledge. The test revealed that IPM-FFS farmers' knowledge reached a satisfactory level, especially with regard to pests and natural enemies identification. The knowledge about ecological functions in the rice field, and negative effects of chemical pest control to the rice ecosystem also reached a satisfactory level. Knowledge increase was mentioned as the main reward of IPM training and implementation, before economic profits as it provided them with more expertise, responsibility and independence in their farm management knowledge about natural enemies *viz.* Spider increased from 21 per cent to 76 per cent, dragonflies from 14 per cent to 46 per cent and beetles went up from zero to 26 per cent, while the knowledge about these natural enemies in non-IPM farmers was 19, zero and six per cent respectively.. The study concluded that the gain in knowledge about arthropod natural enemies in IPM trained farmers was as a result of IPM-FFS. Many farmers had the knowledge that lady bird beetles eat brown plant hoppers and eggs, dragonflies eat brown plant hoppers, caterpillars, moth and eggs. The pre and post knowledge and post knowledge ballot box test showed that gain knowledge of respondents before and after IPM training was with respect to stem borer dead heart appearance (increased to 58 per cent from 44 per cent), stem borer white head appearance (increased to 90 per cent from 73 per cent), leaffolder symptoms (increased to 43 per cent from 16 per cent) leaffolder appearance (increased to 31 per cent from 10 per cent). Braun *et al.* (1995) explained the achievements of FFS in Indonesia, where farmers learnt to train other farmers to promote the use of IPM to control the pests of sweetpotato. One pilot sweetpotato FFS in Indonesia was studied and concluded, that the farmers valued gain in knowledge, mainly on the role and importance of natural enemies in the field. As the FFS programme progressed and the farmers' knowledge increased, more and more farmers felt confident enough to give presentations. Ooi (1998) using three case studies from Indonesia observed that

IPM- FFS programme promoted the action research in the villages. Through this action research farmers designed and conducted experiments in their own fields to establish management techniques for minimizing the damage caused by pests such as white stem borer, rice ear bug and brown plant hopper. Farmers' discoveries led to solutions, thereby not depending on insecticides.

Baca and Gonzalez (1994) studied the extension model and training methods based on the concept of mass education, which were used for dissemination of IPM technology in Nicaragua. The results showed a change in the attitude of the farmers regarding pest management and the use of pesticides.

Mier (1995) in his two years evaluation of Zamorano's IPM programme reported that a year after training, farmers remembered key concepts, such as that insects produce sexually, that many insects are natural enemies of insect pests, and that insecticides kill beneficial insects.

Winarto (1995) explained that IPM programme provide training to agricultural officers and farmers to shift their perspective in pesticides use through 'knowledge transmission' rather than transfer of 'technological packages'. The stimulus to discover unknown phenomena, on the basis of which, farmers improve their knowledge and learning capacity.

Rubia *et al.* (1996) conducted field surveys to assess the knowledge of farmers about white stem borer which had devastated rice crop during 1989-90 outbreaks. Nearly all the farmers (97 per cent) observed white stem borer moths and egg masses (96 per cent). Farmers perceived white stem borer as the key pest and it was felt that extension personnel might need to put more effort in disseminating research information and IPM–FFS reach should be more to avoid unnecessary sprays.

Mangan (1997) reported about the Chinese rice farmers experiences about IPM-FFS. The farmers developed a deeper understanding of how the different components of the rice ecosystem interact which enabled them to make more informed decision about their own field management. Before FFS, only 11 farmers knew the names of those insects that were pests, while 60 per cent of the total sample could not give a specific example of any beneficial in the field. After IPM-FFS all the farmers mentioned spiders as natural enemies. The farmers gained knowledge about different natural enemies and there was 22 per cent increase in knowledge about pest species. During the training the farmers' concept of rice ecosystem and their understanding of beneficial insects in the ecosystem deepened considerably.

Peshin *et al.* (1997) studied the impact of training under FFS in Jammu and reported that the majority of the trained farmers (66 per cent) had higher knowledge level about recommended IPM practices in rice crop, 80 per cent of the trained farmers had knowledge about stem borer and leaffolder, where as only 32 per cent of the trained farmers had knowledge about blast and bacterial leaf blight. Peshin and Kalra (2000a) studied the IPM-FFS programme in rice crop in ten villages of Punjab and reported that IPM farmers possessed higher knowledge level than non-IPM farmers about different IPM practices. There was a significant difference between the knowledge of farmers from IPM villages as compared to non-IPM villages. All the respondents from IPM villages, except one village identified stem borer and leaffolder.

Farmers identified the stem borer damage as white head. The knowledge about planthoppers was not consistent in all the IPM villages which ranged from 13 to 100 per cent. Farmers from IPM villages had also gained knowledge about natural enemies *viz.* spiders and dragonfly (80 and 40 per cent respectively) and were aware of the fact pesticides destroy these natural enemies. The knowledge gain about insect pest and natural enemies was not uniform in all the IPM villages. As the programme scaled up the quality of FFS implementation had decreased. Singh (2002) in his study on IPM-FFS in cotton crop reported that IPM trained farmers had overall higher knowledge level compared to non-IPM farmers. There has been significant knowledge gain of farmers on account of IPM training. Knowledge about natural enemies and different stages of American bollworms was higher in case of trained farmers.

Mostly the studies related to knowledge gain of farmers about IPM practices relate to the farmer field school model of extension followed for implementation of FAO initiated Inter Country Programme (ICP) in rice crop. The training of farmers by way of facilitation rather than conventional top-down approach has resulted in knowledge gain of farmers about insect pests, natural enemies and harmful effects of pesticides.

2.5 Adoption of IPM Practices

The relative success of the IPM extension programmes is ultimately judged on the adoption rate of the IPM systems (or components *there of*) and the improvements in the production associated with this (Dent 1995). Without a unanimously accepted definition there is considerable difficulty in determining the extent to which IPM has been adopted (Norris *et al.*, 2002). The use of truly integrated pest management (based on definition of IPM discussed in chapter 1) is still relatively low as worldwide review of the IPM literature suggests (Kogan and Bajwa 1999). The pest management practices in different agro ecosystems have changed dramatically since the late 1960s in some developed countries (Norris *et al.*, 2002) and since 1980s, in most of the developing countries IPM philosophy has made major contribution in that regard. In 1997, it was estimated that IPM was practised in more than 50,000 communities throughout the world (Kenmore 1997). As such an effort has been made to review the literature to analyse the adoption of IPM practices in different crops.

2.5.1 Adoption of IPM System in Cotton Crop

Thomas *et al.* (1990) studied the adoption of IPM practices among Texas cotton growers. The IPM practices adopted by farmers were scouting (90 per cent), pheromone traps (46 per cent) and conservation of natural enemies (78 per cent). Multiple regression results indicated 32 per cent of variation in the total number of bundled IPM practices. Thirty-six per cent growers adopted all the three practices, 21 per cent adopted one practice and six per cent adopted none of the practices. IPM adoption resulted in higher yields of upland and pima cotton. Lin *et al.* (1994) measured the extent of adoption of IPM for 50 USA crops which included cotton. The study indicated that pest monitoring for all crops and crop rotation for maize crop were among the most widely practiced IPM strategies. The use of beneficial insects was among the least widely practiced. Kogan (1998) in his review of IPM listed the progress of IPM in the USA. The different IPM projects were analysed. The Huffakar project covered six crops–alfalfa, citrus, pines, pome, stone fruit and cotton. The adoption of IPM by growers in these crops led to 40–50 per cent reduction in the use of the more environmentally polluting insecticides within a five years period and 70-80 per cent in 10 years (Huffakar and Smith 1972). Consortium of IPM was second large scale IPM project (1979-85). The coverage of the project was 5.76 million hectares. The main indicators of adoption were the use of scouting, and economic injury level for spray decisions and use of selective pesticides (Frisbie 1985).

Juliana *et al.* (1991) measured the adoption level of IPM practices by cotton growers in five villages of Tamil Nadu (India). Fifty per cent of marginal farmers, nearly half (47.50 per cent) of small farmers and a little more than half (52.50 per cent) of big farmers were found to be medium adopters. About 2 per cent of big farmers were high adopters when compared to small and marginal farmers (22.5 per cent and 5 per cent, respectively). Ramasamy (1994) studied the plant protection practices of cotton growers and adoption of IPM practices, in Tamil Nadu during 1990-91 and revealed that crop rotation and avoiding cultivation of host plants were the two dominant practices being adopted by the farmers. Ragunathan (1995) analysed the data obtained from cotton IPM-FFS conducted all over India. The IPM- FFS had resulted in 30-50 per cent reduction in pesticide consumption in IPM area compared to non-IPM area. The study of FFS in 91 villages of Thanjavur district Tamil Nadu during 1994 season pointed towards farmers adopting agro-ecosystem analysis which increased from 15.8 per cent to 100 per cent, before and after IPM. Decision making in crop management increased from 27.7 per cent to around 100 per cent, and reduction in pesticides was 78 per cent after IPM-FFS. Hugar *et al.* (2001) reported about the adoption of IPM technology in cotton production in Raichur district of Karnataka State (India). The farmers adopting recommended dosage of pesticides was 77.50 per cent, use of pheromone trap 35 per cent, nuclear polyhydrosis virus (NPV) 30 per cent and *Trichogramma* 27 per cent. Lavekar *et al.* (2001) validated the bio-intensive package of IPM on a large scale in district Nanded, Maharashtra (India) during 1998-99. The practices like sanitation, procurement of new seed and synchronized sowing was given more emphasis with the help of FFS. The impact in term of reduction in the use of pesticides cost of plant protection inputs decreased from Rs 3225 to Rs 1545 ha^{-1}.

Ramalho (1994) in his review of cotton IPM: Brazilian perspective reported that cotton growers of Brazil were more concerned about adopting ecologically sound cotton–production tactics since early 1980s and there was significant change in approaches to managing cotton arthropods. An estimated 10 per cent of the farmers in the state of Sao Paulo apply the basic principles of IPM in cotton crop which include cultivation of short-duration varieties, area-wide uniform planting dates, early harvest, stalk destruction, conservation of natural enemies and scouting, but most farmers do not use ETL and scouting for making control decisions.

Sugonyaev (1994) reviewed the progress of IPM in Common Wealth of Independent States (CIS) since 1985. The number of pesticide-treated cotton areas in Turkmenistan and the Murgab Oasis had decreased from 85000 ha in 1970 to 900 ha 1990 and from 125,000 ha in 1980 to 90 ha in 1990, respectively. Over the period, the use of biological control agents in cotton had increased from 54.8 per cent in 1986 to

85.3 per cent in 1990 (Niyazov 1992). In 1994 the mean pesticides consumption was 0.2 kg ha^{-1} throughout Turkmenistan, and 0.1 kg ha^{-1} in the Murgab Oasis. In Tajikistan, the implementation of IPM programmes reduced the number of pesticides applications against cotton bollworm and spider mite three folds (Kovalenkov and Aleshev 1977). This trend was maintained from 1980 through 1989 (Vanyants 1991), as a result the mean load of pesticides decreased 6.8 times per hectare over 20 years.

Poswal and Williamson (1998) reported about the management of whitefly, the most serious pest of cotton in Pakistan with reduced dependence on pesticides through farmer participatory IPM cotton training project. Trainers conducted FFS and it was reported that natural enemy protection were allowed to build up resulting in higher yields.

2.5.2 Adoption of IPM Practices in Other Crops

Adalla and Rola (1986) and Kenmore *et al.* (1987) studied the farmers training in IPM programme and reported that farmers were trained in groups throughout the rice growing season in Philippines. Real plants and insects were used as training materials instead of handouts. Lectures were avoided and technology was greatly simplified. The results were notable as the farmers were able to identify pest problem better and trusted more in their decision making ability. Pesticides used by trained farmers decreased, resulting in reduction of pest control costs. By adopting IPM practices the use and dependence upon pesticides had dramatically reduced in areas where training in IPM had been imparted to rice farmers in Philippines (Galvan and Kenmore 1991).

Bhuiyan *et al.* (1992) conducted a study to examine the adoption of ETL concept by two groups of rice growing farmers in Philippines, one group trained under IPM programme and other not trained under IPM programme. Of trained and untrained farmer, 41 percent and 68 per cent, respectively, applied insecticides following the use of fertilizers. Out of trained farmers, 44 per cent compared to 71 per cent untrained farmers based their decision to apply insecticides after fertilizer application, for offering early protection. From trained farmers, 56 per cent used the concept of ETL compared to 20 per cent untrained farmers, and out of these, 71 per cent and zero per cent of trained and untrained farmers respectively determined ETL before applying insecticides. Chin *et al.* (1992) reported on the implementation of IPM programme in rice in Malaysia, which incorporated pest surveillance, cultivating resistant varieties, cultural control, pesticide management, biological control and training. The farmers successfully adopted IPM, which was due to proper organization, well coordinated research based data, training and strategic campaign.

Kamp *et al.* (1993) reported that in Bangladesh IPM activities were introduced in late 1980s.Most of the farmers having less than half a hectare of land were trained in rice-fish culture and IPM. An ex post facto survey of these areas in 1992-93, revealed that all the farmers had stopped adopting pesticides to control insect pests. The farmers, who could not cultivate fish due to lack of adequate water, poor water-holding capacity of the soil, and risk of flooding, reduced their pesticides applications by 76 per cent with IPM training.

Qadeer and Tomar (1993) presented the results of rice IPM demonstration in Haryana (India) and reported that non-IPM farmers followed the advice of pesticide salesmen. In IPM plots there was 50-70 per cent reduction in pesticide applications compared to non-IPM plots.

van de Fliert (1993) in a case study of IPM-FFS in central Javanese district of Indonesia, reported that out of the trained farmers, 29 per cent selected certified seed, compared to 20 per cent of non-IPM farmers. Only six per cent of trained farmers adopted third or later progenies of seed in contrast to 20 per cent of non- IPM farmers. No significant difference was observed in the adoption of fertilizer practices between IPM and non-IPM farmers, over the four study seasons. The basal fertilizer application was generally not adopted, rather it was disliked, and similarly no change was reported in the water management after IPM-FFS. The change in the adoption of chemical pesticides was observed, as there was a trend of reduced frequency of pesticide sprays. After training, farmers observed pest population before pesticide application. The results of chemical control practices, in two villages revealed significant achievement of IPM training. In IPM villages, frequency of number of spray applications per season decreased from 1.4 to 0.7 to 0.8 to 0.3 before, during, after one season and after two seasons of IPM training, respectively. Number of farmers not using pesticides increased from 26 per cent to 41 per cent to 46 per cent and to 50 per cent, before, during, after one season and after two season of IPM training, respectively. Wiradmadya and Kusmayadi (1996) reported on banning of broad spectrum pesticides for use on rice crop in Indonesia in 1986. This resulted in increase in rice production and reduced use of pesticides. Field schools were found to be successful in educating farmers and increasing the adoption of IPM.

Ragunathan (1995) reported that in 91 villages of Tamil Nadu covered under rice IPM, the farmers' agro-ecosystem analysis increased from 15.8 per cent before IPM to 100 per cent after IPM and decision making in crop management increased from 27.7 per cent to 99.5 per cent and reduction in pesticides was 78 per cent after IPM-FFS. Peshin *et al.* (1997) studied the adoption of IPM practices by farmers trained under IPM-FFS in Jammu, India during 1994 and reported that no farmer had gone for blanket applications of pesticides after training. About 43 per cent farmers had monitored pest-defender ratio (2:1) and 84 per cent of farmers who observed stem borer and leaffolder during 1995 cropping season applied the concept of ETL. Peshin and Kalra (1998) reported the results of study conducted in district Ludhiana (Punjab) to find out the adoption of IPM practices in rice crop after IPM-FFS. No significant difference was observed in the adoption of cultural practices *viz.* fertilizer management, water management, seed treatment, trimming of bunds, dates of transplanting, maintaining desired plant population per square meter and deep summer ploughing between IPM-trained and other farmers. The majority of IPM trained and other farmers had adopted recommended cultural practices. Due to the outbreak of bacterial leaf blight during IPM programme in 1995, 26 per cent of the total area in IPM villages was under resistant/tolerant varieties of rice as compared to four per cent in non-IPM villages during 1996. The manual mechanical practices recommended under IPM programme *viz.* clipping of seedlings before transplanting, rouging-out of infested plants and use of rope method for dislodging leaf-feeders was negligible and concluded that due to mechanization of agriculture in Punjab the possibility of these manual mechanical practices being adopted by the farmers in future was doubtful. The benefits of IPM-FFS were that average frequency of pesticide applications by IPM farmers were 1.36 per season as compared to 2.47 per season by other farmers. Ten per cent of IPM farmers did not spray any insecticides without any adverse effect on the yields.

Fernando (1996) elaborating the achievements of IPM-FFS programme in Sri Lanka, reported that average number of pesticide applications ranged between three to four sprays, before IPM-FFS. After IPM, the farmers reduced the average pesticide applications to just one spray in rice crop. Yongfan and Diyun (1996) analysed the sustained effects of IPM-FFS, on 22 farm households trained in IPM during 1994. Results indicated that adoption of IPM practices resulted in trained farmers spraying their fields during 1995 on an average of 0.3 times more than they did in IPM-FFS 1994. Despite small increase in pesticide applications, the farmers trained in 1994 sprayed the rice crop on an average 1.6 times less then untrained farmers did in 1995, and used 41.63 per cent less pesticides. In Indonesia, in Central Java IPM farmers refused to adopt new credit and input package being promoted in Indonesia, if that was obliged to be connected with pesticides. Farmers organized themselves into block association to manage water and cultural practices more effectively in support of IPM activities. Farmers even created village newspapers and other media to promote IPM adoption. IPM farmers applied pesticides on an average, 1.25 times whereas non-IPM farmers made an average of 3.62 applications per season (Wiradmadya and Kusmayadi 1996). IPM farmers adopted far fewer banned chemicals than untrained farmers.

Musser *et al.* (1986) reported that Georgia groundnut farmer adopted IPM on a limited basis. Although objective data indicated that IPM technology might have been more efficient than conventional pest control strategies. McNamara *et al.* (1991) reported that despite producer and environmental benefits associated with IPM in groundnut, adoption remained low among produces in Georgia, the USA. The study suggested the shift in extension IPM education programmes in the USA, to non traditional audiences, development of education programmes targeted to specific agricultural activities and producer characteristics, which might increase adoption.

Cirio (1992) conducted the study to find out the adoption of IPM in Canino (Italy) area, the results revealed a significant reduction in the quantity of chemical used for pest control and increase in the value of olive crop to non-IPM plots.

IPM is more complex process than relying on regular calendar based applications of pesticides. It requires a level of analytical skills and certain basic training in crop monitoring and ecological principles (Pretty 1995). Where framers have been trained as experts, such as Hondrus (Bentley *et al.*, 1993) and in IPM-FSS in rice there has been adoption of IPM principles. Few IPM programmes have made a lasting impact on farmers' knowledge, attitude or practice (Matteson 1992). In the USA, IPM programmes focusing on better scouting for pest, rotation and other cultural practices have been widely adopted and on many crops IPM is employed on more than 15 per cent of total acreage (NRC 1989). The farmers trained under participatory IPM programmes have widely adopted IPM practices but diffusion of IPM practices to other farmers has not been reported.

2.6 Economic Rewards of IPM Programmes

Economics is the key to adoption of new crop production system (Lacewell and Taylor 1980). Diffusion of innovation theory listed different characteristics (attributes) of innovation which affect its rate of adoption, and one of these relative advantages in terms of economic benefit is very important for adoption of a particular crop production system. New crop production system must increase the farmers profit or he has no incentive to adjust from traditional crop production practices. The increase in profit for a producer due to use of IPM can be increased yield, reduced costs, or reduced yield and reduced cost of production where the cost reduction is greater than the value of yield reduction (Lacewell and Taylor 1980). The economic proof of IPM's benefit, though shown in many projects in India and rest of the world, but it has to be brought directly to farmer or farmers.

Lacewell and Taylor (1980) reviewed and analysed some IPM programmes in the USA prior to 1980 by statistical analysis. In 1979 the short- season cotton IPM programmes resulted in an $ 11 million increase in producer profits. The yield increase was 30 per cent, reduction in pesticide use was 27 per cent, energy use reduced by 33 per cent and profit increased 646 per cent (Sprott *et al.*, 1976). The impact of such programme in lower Rio Grande Valley indicated that for 1976-78, the yield for short-season and conventional cotton production practices were not significantly different. Short season system was adopted by all producers and no conventional producers existed (Collins *et al.*, 1979). In West Texas a short-season IPM cotton production system was developed that shifted to the deterministic varieties, reduced fertilizer and in turn irrigation thereby reducing foliage insect pest problem (Condra *et al.*, 1975). In this IPM programme, yield was reduced; however cost of production was reduced dramatically. Yield reduction was 11 per cent, pesticide use reduced by 71 per cent, nitrogen fertilizer use reduced by 76 per cent, irrigation water usage decreased by 25 per cent, cost of production per acre reduced by 46 per cent and profit increased from a negative $ 105 to a positive $81.50 per acre (Condra *et al.*, 1975).

Texas extension service state wide cotton IPM programmes started in 1972 in Texas. The comparison was made on the basis of programme participants and non-participants and acres. Energy savings with IPM programme 0.5 gallons in 1973 and 2 gallons in 1974 of fossil fuel equivalent. The social benefits in terms of benefit cost ratio was $7.33 returns in 1973 and $4.60 returns in 1974 for each dollar of programme cost, including programme cost allocated to the farmers (Lacewell *et al.*, 1974). Lacewell and Taylor (1978) worked out the benefit of extension service IPM programmes implemented throughout the south of the USA. Three of the four programmes showed no yield change but a substantial reduction in number of pesticide application. In Mississippi 28,233, Albama 38,759 and North Carolina 76,000 acres were covered under cotton IPM programmes. The pesticides application reduced by 2.8, 20.0, 6.0 in Mississippi, Albama and Arkansas respectively. Reduction in cost per acre was small in Mississippi ($ 3.50) and Alabama ($ 3.00), though over several millions acres the impact could be substantial. But the incentives to farmers in many cases were not sufficient to attract producers.

There were a number of IPM programmes for crops in the USA. IPM programme for grain sorghum in Oklahoma, which indicated an increase in producer profit of about $ 3 per acre (Salkin *et al.*, 1976). An IPM programme for alfalfa was estimated to save farmers in eastern USA about $ 44 million annually (Zavaleta and Ruesink 1980). Biological control of parlatoria scale olives in California saved the industry $ 1 millions annually in control costs and IPM programme in apples reduced treatment by 30 per cent (Huffakar and Lacewell 1978).

Lacewell and Taylor (1980) while analysing different IPM programmes in cotton concluded that short season cotton production system for south Texas increased the yields by 30 per cent, reduced costs by 33 per cent and profit increase from $ 12 per cent acre to $ 103 per cent acre resulted in an $ 11 million increase in producer profit and total economic impact of $ 94 millions. The other IPM programmes prior to 1980 showed a benefit-cost ratio of 7.3 and 4.6 but at same time IPM programme throughout the USA were estimated to cause decline in farmer profit by $ 593 million.

National Research Council (NRC), USA reported about the impact of IPM programmes in nine crop commodities. Even through the number of pesticide applications had increased, but the volume of pesticide applied had reduced, and the yields increased by 110-130 per cent with an annual saving of $ 578 million. Volume of pesticides has declined because of precise timing and the use of more specific products (NRC 1989).

However, cotton insect pest management in high input production region of the USA was still largely based on the use of chemical insecticides (Luttrell 1994) the number of annual insecticide applications on US cotton during 1961-77 averaged 5.7 (Ridgway *et al.*, 1993) and in 1992 approximately 5.8 (Luttrell 1994). But at the same time in certain regions the average number of insecticide applications decreased or stabilized from 1988 to 1992. The number of insecticide applications in cotton during 1988 in West USA and South east USA were 4.3 and 9.3 respectively, which decreased to 2.9 and 6.0 respectively (Head 1993).

The economic benefits of cotton IPM programme in Australia since the publication of IPM guidelines in 1999 and the initiation of the extension efforts to promote IPM has showed downward trend in the number of insecticide spray applications and increased trend towards more selective insecticides especially in *Bt* cotton (Wilson *et al.*, 2004). It is reported that overall amount of active ingredient (a.i) per ha showed a marked drop since the emphasis on IPM. From more than 14 insecticide sprays (conventional) and 8 insecticide sprays (*Bt* cotton) in 1998-99 these declined to less than 8 and less than 4 in 2001-2, respectively. The consumption of average amount of insecticides (a.i) decreased from about 11 kg ha^{-1} (conventional cotton) and about 6 kg ha^{-1} (*Bt* cotton) in 1998-99 to around 4 kg ha^{-1} (conventional cotton) and less than 2 kg ha^{-1} (*Bt* cotton) in 2001-02 due to adoption of IPM and gross margins were same or higher in case of IPM approach than non-IPM approach, which means that net returns in IPM approach were higher (Wilson *et al.*, 2004).

In Brazil in late 1970s, the resistance of *Heliothis* to organophosphates was a major problem. At that time cotton received 15-20 insecticides applications per season. The launching of IPM in 1979, helped significantly to reduce insecticide applications on an average to six sprays per season (Seganmullar 2000).Cotton growers in Brazil have implemented IPM programmes on a large scale in early 1980s and thereby optimizing profits through lower production cost for the same level of yield (Pimentel and Banderia 1981). But in 1983, the introduction of *Anthonomas grandis* to Brazil, the number of insecticide applications again rose to 10-12 applications per season, but after local behaviour pattern was established for new pest, and IPM adapted accordingly, the number of applications decreased again to an average of 8 per season (Seganmullar 2000). IPM has produced excellent economic, social and ecological results in Brazil (Cruz 1991).

The adoption of IPM practices in cotton in Common Wealth of Independent States (CIS, formerly USSR) has produced excellent results in terms of economic benefits. The yield of raw cotton in Murgab Oasis increased, by 0.26 tons ha^{-1} from 1981 to 1990. In addition, at the 2500 ha cooperative farm Tezedurmush, introduction of IPM principles in the 1970s produced savings of 20,000 rubles annually from reduction in pesticide expenses alone and additional production of raw cotton over 5 years was 1000 tons (Niyazov 1992, Vanyants 1991). In Tajikistan, implementation of IPM programmes reduced the pesticide use from 35.2 kg ha^{-1} to 5.2 kg ha^{-1} over 20 years. Economically, yield of raw cotton increased about 30 per cent and returns on expenditure on cotton protection was 13-14 times higher in IPM areas with additional income of 880-896 rubles ha^{-1} in the CIS states (Vanyants 1991).

In Sudan, the programme for development and application of IPM in cotton resulted in reduction in the cost of plant protection to total cost of production from 33.3 per cent in 1985-86 to 19.3 per cent in 1988-89, with the reduction of average number of pesticide applications from 8.5 to 5.3 for the same period. The yield of seed cotton increased from 3.4 q ha^{-1} to 4.1 q ha^{-1} (long staple), and 5.5q ha^{-1} to 6.5 q ha^{-1} (medium staple) from 1985-86 to 1988-89 (Oudejans 1991).

2.6.1 Economic Benefits of IPM Programmes in Asia

In India, Operational Research Project (ORP) for cotton and rice was started in 1975. The results of the project showed significant economic benefits. In Punjab, a total number of 15 villages in the main cotton growing districts (Faridkot, Bathinda and Ferozepur) were covered during the period of 15 years. Adoption of short duration varieties, jassid tolerant varieties and other IPM practices resulted in 73.7 and 12.4 per cent reduction in number of insecticide sprays for control of sucking pests and bollworms, respectively (Dhaliwal *et al.*, 1992). In Tamil Nadu the ORP farmers obtained 23.2 per cent higher yields and 31.7 per cent higher net income. The number of insecticide applications in IPM area was 6.3 compared to 10.7 in non-IPM area, and the average seed cotton yield in IPM area was 26.7 as against 22.3 q ha^{-1} in non-IPM area (Simwat 1994).

CAB International Biosciences trials were carried out on small holder cotton farmers in Punjab and Maharashtra. Twelve farmers participated at each location to validate the IPM technology. In Punjab IPM farmers made 60 per cent fewer sprays of insecticide and the cost of cultivation of IPM plots were 13 per cent less (Rs 495 ha^{-1}) than the cost of insecticide application for non-IPM plots. In Maharashtra financial benefits of IPM practitioners were more marked than in Punjab although yields in

IPM plots were only 10 per cent higher. The synthetic insecticide applications were lowered by 96 per cent (CABI 2000).

The state departments of agriculture in different states have been running IPM programmes funded by Directorate of Plant Protection Quarantine and Storage, Government of India, with active cooperation of Central Integrated Pest Management Centers set up in each state but the overall economic impact of these programme are not well documented except for the annual reports of these programmes. So impact of IPM technology in cotton over a period of time is not well documented. Same is the case with China and Pakistan (Luttrell *et al.*, 1994).

Insecticide resistance management (IRM) with major focus on rationalizing insecticide use was started initially in four states of India. Results in 1999 were very encouraging as average yield increases varied between the participants states from 17-70 per cent, plant protection costs reduced by 21-88 per cent and net profit increased from $ 40 to 226 ha^{-1}. The average insecticide use reduction of the 3500 farmers was 40-60 per cent with a modest 12 per cent increase in yield and 48 per cent increase in profitability across the ten states (Russell *et al.*, 2003).

2.6.2 Farmer Field School Programme in Asia

The success of what is now seen as the 'classical' farmer field school (FFS) programme on rice in Asia (Pontius *et al.*, 2002) has encouraged the testing and adoption of the methodology in cotton. In India, cotton-FFS were run in five states, yield increase were ranging from 21 to 27 per cent and pesticide reduction from 30 to 50 per cent (Dhaliwal *et al.*, 1998). In 2002 cotton season, some 800 FFS were run (Anonymous 2003b), in Bangladesh, China, India, Pakistan, Philippines and Vietnam per cent of the world's cotton (Eveleens 2000). Results on experimental plots in 2002 in China showed that FFS-IPM farmers in conventional cotton made net profit of $ 1,150 ha^{-1} versus $ 520 ha^{-1} for normal farmer practice areas (Anonymous 2003b). In India cotton FFS prior to 1995 showed 20-26 per cent increase in yield and 30-50 per cent reduction in pesticide consumption in IPM areas compared to non-IPM areas (Ragunathan 1995).

The economic benefit of IPM-FFS in rice crop were reported by van de Fliert (1993), Ragunathan (1995), Fernando (1996), Hung and Nhat (1996), Sattar and Shafiquddin (1996) and Peshin and Kalra (1998). In India, data from 688 rice–FFS showed an increase in yield up to 34 per cent and 50-100 per cent decline in the use of pesticides in IPM areas as compared to non-IPM areas. A case study in 91 villages in Thanjavir district, Tamil Nadu during 1994 season showed a reduction in pesticide was 78 per cent after IPM-FFS. The productivity of these FFS village increased from 47.27 q ha^{-1} to 53.059 q ha^{-1} (Ragunathan 1995). In district Ludhiana, Punjab an ex post-facto study of IPM-FFS village in 1996-97 revealed that average frequency of pesticide application before, during and after IPM training in six villages was 1.88, 1.64 and 1.5 per season respectively. The average yields per unit area in these IPM villages were 63.13 q ha^{-1} compared to 52.28 q ha^{-1} in non-IPM area. The average frequency of pesticide application in 10 IPM village during 1996 (after IPM programme) was 1.36 compared to 2.47 in non-IPM villages and IPM farmers insecticides expenditure was Rs. 364 ha^{-1} compared to Rs. 1063 ha^{-1} by non-IPM farmers (Peshin and Kalra 1998).

In Bangladesh, in 68 IPM-FFS there was 90 per cent reduction in pesticide use by the farmers after IPM training. Before training farmers were applying at an average of 1.65 number of pesticide applications, where as after training average pesticide applications were reduced to 0.19, which showed a reduction of 88 per cent. By adoption of IPM practices *viz.* high yielding varieties, improved cultivation practices and proper weed management, the yield increased by 17 per cent (Sattar and Shafiquddin 1996). In Cambodia the results of five FFS in rice showed that farmers saved \$ 38 ha^{-1} by not applying insecticides without adversely affecting yield per ha (Nody and Chhay 1996).

In China, results of IPM-FFS in 1995 indicated that adoption of IPM practices resulted in trained farmers spraying their fields at an average of 0.3 times more than they did during IPM-FFS in 1994. It resulted in 5.5 per cent increase in pesticide used one year after training as well as 7.4 per cent increase in 1995 by 1.4 per cent over 1994. Despite this small increase in pesticide applications, the farmers trained in 1994 sprayed their rice crop on an average 1.6 times less than untrained farmers did in 1995 and used 41.63 per cent less pesticides. Trained farmers saved on an average RMB 4.98 per mu in 1995 season compared to untrained farmers. Average yield of trained was 38.21 kg mu^{-1} higher than untrained farmers in 1995, and net income of trained farmers was RMB 51.22 per mu^{-1} higher than that of untrained farmers (Yogfan and Diyun 1996).

The benefits of IPM-FFS programme in Sri Lanka were that average number of pesticide applications ranged between three to four sprays and one spray before and after IPM-FFS, respectively, which resulted in reducing the cost of pesticides by 50-70 per cent and increase in yield by 9-25 per cent (Fernando 1996).

In Indonesia the average insect control cost per ha decreased during after IPM-FFS in rice and the average yield increased from 5.38 to 5.77 to 3.70 to 6.38 tons ha^{-1} before, during, after one season and after two seasons of IPM training, respectively, and reduced expenditure on insect control compared to non-IPM farmers (van de Fliert 1993).

Based on FAO country briefs of ICP programme for IPM in Asia 1996, the adoption of IPM practices in rice crop resulted in yield increase per unit area by 2.46–25 per cent and reduction in pesticides cost by 27-86 per cent (Peshin 2002). The long-term impact analysis of IPM programmes in India to quantify the reduction in use of pesticides, if any, and the economic benefits of these programmes is must, as is being done in most of the developed countries like the USA and developing countries like Indonesia. In India, the empirical evidence of the benefits of IPM programmes after the intervention in the form of projects/programmes withdrawl is needed to justify the implementation of different programmes related to IPM.

2.7 Adoptability and Constraints

When pest control was simply a process of spraying pesticides, many farmers carried out their own decision-making duties, often obtaining the information about pesticide use from pesticide sales representatives, pesticide suppliers (Norris *et al.*, 2002) or in some cases from the extension workers of state agricultural departments.

Pesticide retailers/pesticide agents of the pesticide industry played a major role in the decision making process regarding chemical pest control (Peshin and Kalra 1998). Due to complexities of carrying out IPM at field level the constraints faced by the farmers are numerous, some of these can be generalized and some are location specific. The sustainable practices like IPM are generally complex because of their interactive nature. The decision to adopt IPM practices, therefore, must be compatible with the existing production system and overcome the resistance of change farmers possess (Rolling 1988). The attributes of a particular IPM practice are also responsible in its adoption. Attributes in terms of perceived or actual benefits; relative advantage (Drost *et al.*, 1996), compatibility with existing production (pest management) practices (Drost *et al.*, 1996), complexity of carrying out IPM at farmers' level (Norris *et al.*, 2002) effect/limit the adoption of IPM. The constraints/barriers in the adoption of IPM have been reported from around the world.

Musser *et al.* (1986) reported that in Georgia (USA), ground nut farmers who did not adopt IPM had unfavourable views pertaining to the consequences of employing IPM compared to adopters. The belief of non-adopters was reported to be the constraints in the adoption of IPM.

Bentley and Andrews (1991) reported about the slug control techniques including 'trash trap'. Despite some Handuran farmers' interest and experimentation, however adoption rates were low. The constraint was because of the non-compatibility of the practices in maize crop. The sharp edges of maize leaves scratched people's faces, as they walked through the corn under the August sun. It was suggested that scientists should understand the limitations in adoption of a particular practice.

Oudejans (1991) reported that small scale farmers in Asia had little knowledge either about pest or disease problems, or about selecting the right product, dosage rate and training for controlling individual pests, or about safe handling which hampered farmers to take decision about control measures. Small scale farmers had great difference in terms of gained knowledge and skill, and access to product and equipment between large scale commercial enterprises.

Bernardo (1993) listed the possible constraints to a wider adoption of IPM in crop protection. The constraints included, the unacceptability, of the recommended pest population, monitoring procedure (counting insects on sample plants), to assess the need for control tactics, the limited availability of biological control agents and inadequacy of IPM programmes for multiple pest problem.

van de Fliert (1993) reported about farmers trained under IPM programmes being doubtful about the IPM concept and therefore continued to apply pesticides in order to feel at ease. Though in the end their doubts were removed, when despite pesticide application, pest could not be controlled resulting in greater economic loss. The constraint observed was that IPM farmers, who had gained knowledge and acquired skills, did not feel competent to react adequately to pest occurrence. In the two IPM villages intensively studied, it was observed that much attention was not paid to control measures. In the case of stem borer out break during 1990-91, IPM farmers' observation skills about stem borer egg masses were not sufficient enough. The constraints in adopting proper control measures, the season after IPM training for controlling stem borer outbreak was mainly due to training curriculum. As the field school graduates emphasized field monitoring instead of agro-ecosystem analysis, informed decision making and adequate pest management measures which made the sustainability of IPM practices doubtful. Role of extension workers as an IPM trainer became dubious in certain cases, as they took advantage of their position and strongly promoted inputs or even sold them from their houses to receive commission for those services, often as indispensable extra source of income. The study also revealed that one extension worker was observed to recommend pesticide in IPM-FFS (incorrectly). The conflict between sustainable IPM practices and input packages given to farmers including pesticides was observed. The counteracting forces, such as extension workers promoting pesticides or enforcing pesticides purchase, were the constraints. These constraints prevented these farmers to adopt IPM committedly.

Matteson *et al.* (1994) concluded on the basis of evaluation of IPM programmes in the Philippines that IPM training deteriorated after being taken out of the hands of 'enthusiastic master trainers', which was one of the constraints in the adoption of IPM by farmers. Farmers did not adopt numerical sampling (ETL), even through some farmers learned sampling, but refused to adopt it, claiming that they could assess pest population visually, without counting or sampling for pests. Merchant and Teetas (1994) reported about the uncertainty in the farmers about their ability to accurately sample insects as the reason for not using ETL by some farmers. Even pest advisors in Texas (USA) showed distrust of ETL for not using these.

Murray (1994) concluded that even though IPM programmes in Latin America showed considerable promise, but the constraints for its success were basically due to its top down approach and its heavy reliance on high qualified personnels, governments found hard to fund. The constants in adoption of IPM by farmers were due to those chemical companies and their agents who lured farmers back into the pesticide mold.

Kenmore *et al.* (1995) stated that top down approach promoted by traditional extension generated little interest in farmers and contributed to poor adoption of IPM. Lack of knowledge led farmers to follow suggestions that pesticides were necessary in rice fields. The trend was changed with the introduction of non-formal approach of educating farmers in FFS.

Winarto (1995) reported that even though 1989 national IPM programme in Indonesia was a case of break through in national policy to enhance the ecological balance by conserving natural enemies and diminishing the indiscriminate use of pesticide in the protection of food crops. IPM knowledge and training provided initial understanding of and stimulus to discover unknown phenomena, but the constraints largely were due to the inflexibility of subsidized scheme of inputs and the inadequate explanation provided to the farmers by extension services.

Bentley and Andrews (1996) reflected on their experiences with promoting the use of IPM in Latin American farming. The main constraints to adoption of IPM were reported to be, insufficient technical knowledge, a weak public sector, inappropriate credit and subsidy schemes, influential agrochemical companies, agro ecosystem complexity, and language barriers between farmers, scientist and extension agents. It was further added that no quantitative strategy of sampling (ETL) would be adopted or accepted by small holders, who lacked the arithmetic and time to count several species in their small fields every few days.

Drost *et al.* (1996) reported that transition problem often limit the adoption of sustainable farming practices. Perception problem often limit the adoption of such practices. Farmers in Utah (USA) developed and adopted cultural practices with which they were comfortable. The barriers in adoption of sustainable practices including IPM were financial, maintaining profits; lack of knowledge and federal farm programmes.

Peshin and Kalra (2002) reported about the constraints in the adoption of IPM practices, which included lack of knowledge about the benefits of proper fertilizer management in rice crop as the topics of fertilizer management were not covered comprehensively during IPM programmes in Punjab. The manual-mechanical practices recommended under IPM programmes were either not explained in detail as farmers lacked proper knowledge about these and were not sure of the benefits of these practices. Manual mechanical practices being labour intensive and lacking instant observable results, the adoption of these was negligible. The contradictory technology disseminated (IPM versus conventional recommendations) was also a barrier in the adoption of IPM practices.

Malone *et al.* (2004) reported that in Virginia (USA) the barriers in the adoption of IPM practices were, lack of time for farmers to personally scout their fields as it was not economically feasible to hire independent crop consultant. The agricultural suppliers and chemical dealers often scout without charge and out number Cooperative Extension Agriculture and natural resource (ANR) agents in scouting fields, which resulted in non-adoption of some of the IPM practices. The IPM resources like Virginia Tech Website were infrequently used due to lack of awareness and limited computer or internet access. The lack of better educated scouts (Sorensen 1993) was also a constraint in IPM practices not being popular.

The constraints in adoption have been in terms of inappropriateness of technology, economic implications, non-availability of appropriate information, lack of knowledge and skills for applying the IPM in the fields, faulty methodology for dissemination of IPM, vast network of chemical industry to lure farmers for using pesticides. Appropriateness of technology in terms of it being less complex and compatible with farming system positively affects its adoption. Due to complexities of carrying out certain IPM practices, it has been difficult for farmers in carrying out IPM practices like ETL (Goodell 1984, van de Fliert 1993, Escalada and Heong 1994, Matteson *et al.*, 1994, Malone *et al.*, 2004). The compatibility of an IPM practice also plays role in its adoption. If IPM practice is not compatible like *'trash trap'* in maize (Bentley and Andrews 1991) it is a limitation in its adoption. Economic returns/ implications of IPM need to be demonstrated to the farmers so that the farmer learns that even buying information and advice can be more profitable than buying chemicals (Lacewell and Taylor 1980). Growers perceived that IPM practices are more risky than conventional pest management (Norris *et al.*, 2002) so the risk associated must be decreased to make farmers sure of its economic viability.

Dissemination of IPM technology related information in top down approach is also a constraint in many developing countries (Kenmore *et al.*, 1995) and lack of proper knowledge about different aspects of IPM, agro-ecosystem analysis and not acquiring required skills for its use acted as barriers (van de Fliert 1993, Merchant and Teetas1994). Vast network of pesticide companies in developed and developing world also lured back the IPM practitioners. The company agents scouting farmers' fields and assisting them act as a barrier for IPM adoption. Counteracting forces even in public extension services confuse the farmers, and the lack of commitment of extension agencies to IPM limit the spread and adoption of IPM (van de Fliert 1993). Lack of master trainers also acted as an obstacle in the adoption of IPM (Matteson *et al.*, 1994).

The constraints for different agricultural systems can vary as in most of the Latin American countries there is no public service extension, so the farmers are more dependent on agents of chemical industry for information. In the USA, the constraints are in terms of IPM adoption being often more expensive than conventional pesticides based management, due to increased need for population assessment and record keeping and where it meets economic interest of growers adoption is high. In developing countries counteracting approaches, lack of proper dissemination of technology in a participatory mode are the barriers in adoption of IPM. For different crops also the constraints differ. The lack of knowledge in terms of comprehension and its applications, and lack of skills to use complex practices are the universal constrains reported in numerous studies.

Chapter 3

EVALUATION RESEARCH: A Theoretical View

A social researcher, which includes the researcher from the field of extension education, can seldom conduct research work without any concern for theory. In many cases, the researcher has some specific theoretical ideas in mind at the commencement of his research project and at other times researcher from working with his data concludes certain things. Whether researcher thinks about certain ideas, or makes certain hypotheses before or after concluding research, the researcher is attempting to construct a theory (Lin 1976). In other terms researcher is trying to generalize certain relationship between variables which may be true not only for the selected respondents, but for people in larger or other social systems. Theory, in its scientific sense, serves one important purpose: *It explains the relationship among observed activities* (Lin 1976). *Theory is a set of interrelated constructs, definitions and propositions that present a systematic view of phenomena by specifying relations among variables, with the purpose of explaining and predicting the phenomena* (Kerlinger 1973). Thus theory refers to relationship, among facts and ordering them in a meaningful way.

Theory provides major orientation to research by helping to define the type of data to be collected and theory emerges from facts (data) collected during research. Theory and facts are not opposed but intervened, theory is not speculation and scientists are concerned with theory as well as facts. Keeping in view the research design, content areas and objectives of the study, theoretical orientation is presented under the following heading:

3.1 Evaluation research: A theoretical view

3.2 Types and forms of programme evaluation

3.3 Methodologies for evaluation of IPM programmes

3.4 Theoretical framework of the study

3.1 Evaluation Research: A Theoretical View

There is no uniformly accepted definition of what constitutes evaluation research (Kosloski 2000). The field of evaluation research can be defined as "the use of scientific methods to measure the implementation and outcome of programmes for decision-making purposes" (Rutman 1984). A broader and more widely accepted definition is "the systematic application of social research procedures for assessing the conceptualisation, design, implementation, and utilization of social intervention programmes" (Rossi and Freeman 1993). A much border definition was offered by Scriven (1991) who defined evaluation as "the process of determining the merit, worth and value of things." It does not limit to a social programme or specific type of intervention but encompasses everything. The object can be a programme, a project, a product, a policy, or a one-of event. Evaluation as such should begin with a close examination of the purpose of the evaluation and clear understanding of the programme and target clientele. It is not until the purpose and main evaluation questions have been agreed upon that the selection of appropriate methods should be considered (Petheram 1998).

3.1.1 Programme Evaluation

Programme evaluation is a form of applied social science which intends to assist in improving the quality of human services (Posavac and Carey 1989). Programme evaluation evolved primarily in the USA and became a semi-professional[1] discipline around 1960s and expanded during the 1970s (Shadish *et al.*, 1991, Petheram 1998).

The theory and practices of programme evaluation that emerged during this time had roots in earlier work done by Ralph Tyler in the period between 1930 and 1945 in education (Tyler 1967), Lewin (1948) in social psychology and Lazersfeld and Rosenburg (1955) in sociology. The interventionist role that the US federal government took on social policy during 1960s and the increasing interest in evaluation research as an academic pursuit (Shadish *et al.*, 1995).

During the early part of twentieth century systemisation, standardisation and efficiency in the field of programme evaluation in education was observed (Petheram 1998). Tyler came to have enormous effects upon the field of evaluation and coined the term 'educational evaluation'. Tyler's approach calls for the measurement of behaviorally defined objectives, it concentrates on learning outcome instead of organizational and teaching input (Madaus *et al.*, 1991). Around 1960, the US government invested large sums of money towards programmes in education, income maintenance, housing and health. The vast expenditure on these increased the demand for evaluation (Bell 1983). By the late 1960s in the USA and internationally, evaluation research had become a growth industry (*Wall Street Journal*, cited by Rossi *et al.*, 1979).

Programme evaluation is semi-professional as it shares certain attributes with other professions and differs from purely academic specialties. It is not fully professionalised as it has no licensure Laws (Shadish *et al.*, 1995).

As evaluation strategies and methods matured, evaluation theory developed towards one focusing on methodology, but in a broad context (Shadish *et al.*, 1991). Around 1973, programme evaluation began to emerge as a semi-professional discipline (Petheram 1998). Around late 1970s and early 1980s, evaluators realised that the techniques of evaluation must:

1. Serve the need of the clients,
2. Address the central value issues,
3. Deal with situational realities,
4. Meet the requirements of probity, and satisfy the needs for veracity (Madaus *et al.*, 1991).

3.1.2 Evaluation Theories and Techniques

Evaluation theory comprises the many possible decisions about the shape, conduct and effect of evaluation, and is, therefore, far from focusing on methods only (van de Fliert 1993). The discussion on methodology include the philosophy of science, public policy, validity, and utilization and the emphasis must be on making the methodology fit the needs of the society, its institution and its citizens, rather than the reverse (Kalpan 1994). In 1988, Brandle's message in an address to professional evaluators was, "No demonstrable relationship exists between programme funding levels and impact, that is, between the inputs and outputs, more money spent does not mean higher quality or greater results until systematic accountability is built into government, no management and improvements will do the job."

3.1.2.1 Rational Design of Evaluation

Evaluation research drew heavily on the experimental model of Donald Campbell, which is based on quasi- experimental research designs (Campbell and Stanley 1963, Cook and Campbell 1979). Much of his work was actually de-emphasized quasi-experimentation in favour of experiments (Shadish *et al.*, 1991). Quasi-experiments frequently lead to ambiguous causal inferences, sometimes with dire consequences (Campbell and Erlebacher 1970). As a result, Campbell is credited proposing a rational model in which programme is first evaluated using rigorous social science methods, such as experiments, when possible, and then a report is issued to a decision maker who acts on the findings (Kosloski 2000). Campbell's emphasis was on internal validity which was consistent with his focus on experiments, as these are particularly useful in examining causal relationship. But according to Cronbach (1982), even though experiments have high internal validity, they tend to be weak in external validity and external validity is of greatest utility in evaluation studies. Internal validity refers to whether the innovation or treatment has an effect, in contrast, external validity addresses the issue of generalizability of effects, specifically, "To what populations, settings, treatments, variables and measurement variables can this effect be generalized" (Campbell and Stanley 1963). The rational model of evaluation research based on rigorous social research procedure is a flawed model because there are no reliable methods for generalizing beyond the factors that have been studied in the first place and it is generalized rather than the specific findings in which evaluators are interested (Cronbach 1982). Cronbach viewed evaluations as more of an art than a scientific enterprise.

3.1.2.2. Meta-analytic Design of Evaluation

Evaluations syntheses represent meta-analytic techniques of evaluation, in which research results from numerous independent evaluation studies are first converted to a common metric and then aggregated using a variety of statistical techniques (Kosloski 2000). The product provides a meaningful summary of the collective results of many individual studies. Meta-analysis has helped to resolve the controversy over the priority of internal versus external validity. "If studies with rigorous designs are used, results will be internally valid and by drawing on findings from many different samples, in many different settings, using many different outcome measures the robustness of findings and generalizability can be evaluated as well" (Kosloski 2000). Meta-analysis has many strengths including increased power relative to individual studies to detect treatment effects, the results are limited by the quality of the original research studies. The major draw back of meta-analysis is repeating or failing to compensate for the limitations inherent in the original research on which the synthesis is based (Figueredo 1993). This is because, many evaluation use non-experimental designs, these methodological limitations can be considerable, although these potentially exist in experiments as well as large portion of experiments suffering from low-external validity (Kosloski 2000).

3.1.2.3 Critical Multiplism Model

Research syntheses of experimental and non-experimental studies is referred to as critical multiplism (Shadish 1993) and based on Campbell and Fiske's (1959) notion of multiple operationalism. 'Multiplism' refers to the fact that there are multiple ways of proceeding in any research endeavour, with no single way being uniformly superior to all others (Kosloski 2000). That every study will involve specific operationalization of cause and effects that necessarily under represent the potential range of relevant components in the presumed causal process while introducing irrelevancies unique to a particular study (Cook 1993). The term 'critical' refers to the attempt to identify biases in the research approach chosen. The critical multiplism is to synthesise the results of studies that are heterogeneous with respect to source of bias and to avoid any constant biases. In this manner, meta-analytic techniques can be used to implement critical multiplistic ideas (Kosloski 2000), thereby our confidence in the generalizability of evaluation finding.

3.1.2.4 Participatory Evaluation

The purpose of this technique is for evaluators and clientele to initiate a critical reflection process focused on their own activities. The model assumes a democratic participatory process along with autonomy on the part of educators and learners at the local level (Brunner and Guzman 1989, Green 1988). This is a form of evaluation, which is usually called 'participatory action research' (Deshler 1998). This technique of evaluation works on the premise that farmers/people at the receiving end are ultimately the best judges to study the impact–whether benefits have been produced or not (Uphoff 1989, 1992). Being included in planning, implementation, and evaluation will show farmers that they are regarded as responsible, capable individuals and not simply passive beneficiaries or a target group (Deshler 1998), and according to Deshler participatory evaluation is self educating, without which sustainable agriculture is unlikely. Kosloski (2000) considers participatory evaluation as a controversial approach to evaluation research. Participatory evaluation is inconsistent with notion that investigator should remain detached from the object of investigations in order to remain objective and impartial.

3.2 Types and Forms of Programme Evaluation

The broad, overall purpose of programme evaluation is to provide feedback to human actors and specific sub-purposes always need to be defined for specific evaluation studies (van de Fliert 1993). Commonly two main types of evaluations research are distinguished, called *formative* and *summative* evaluation (Scriven 1980, Neuman 2000). Another division between types of evaluation is based on the kind of questions asked about programme. These are *need, process, outcome* and *efficiency* evaluation (Posavac and Carey 1997). Owen (1993) classified evaluation by categorising forms of evaluation by purpose which are:

- ✰ Form 1: Evaluation for impact assessment
- ✰ Form 2: Evaluation for programme management
- ✰ Form 3: Process evaluation
- ✰ Form 4: Evaluation for design clarification
- ✰ Form 5: Evaluation for programme development

But all evaluation theorists do not agree that evaluation for design classification and evaluation for programme development are evaluation at all. The evaluation for programme development is largely called in agricultural extension as need or situation analysis rather than evaluation (Petheram 1998).

Petheram (1998) concluded that overall, programme evaluation is now more relevant to agricultural extension in its many forms than it has ever been, and some of the approaches, trends as enlisted by Petheram are given in Figure 3.1.

3.2.1 Formative and Summative Evaluation

Scriven (1967) was the first to write about differences between formative and summative evaluation. These terms have become almost universally accepted in the field of evaluation. 'Formative evaluation' is conducted to provide programme staff judgment useful in improving the programme. 'Summative evaluation' is generally conducted after completion of the programme (or when programme has stabilised and for the benefit of external audience or decision-maker). The main difference is that the aim of summative evaluation is to report on the programme, whereas a formative evaluation reports to the programme (Scriven 1991).

3.2.2 Process and Outcome Evaluation

The division between types of evaluation based on kind of questions asked also differentiates between moments of evaluation in programme design and implementation (Posavac and Carey 1989).

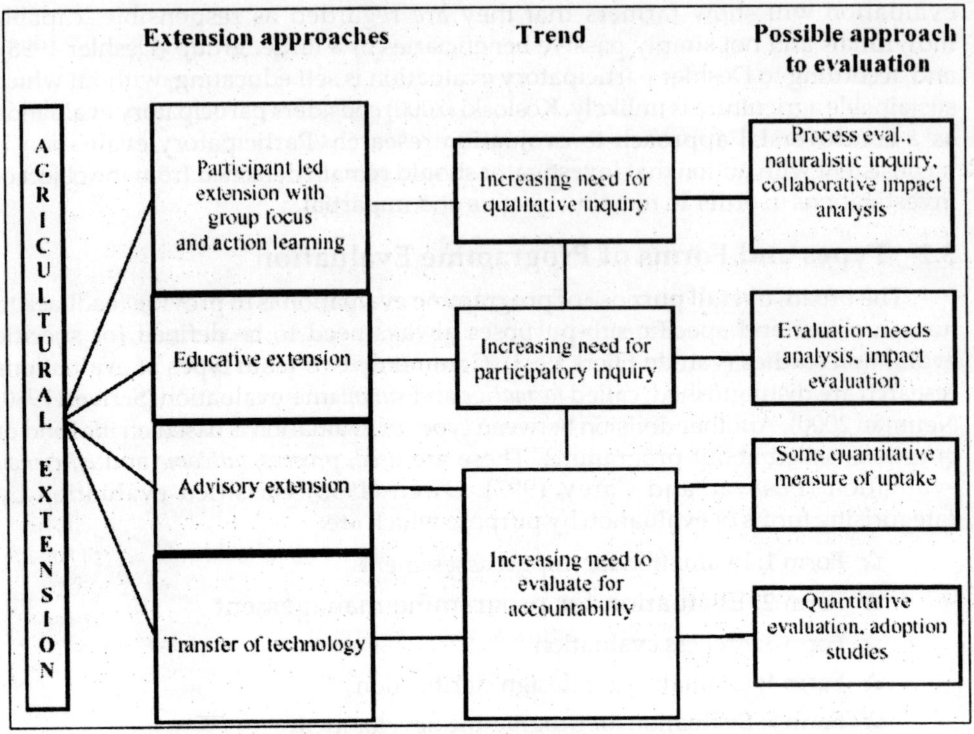

Figure 3.1: Some Trends in Focus of Agricultural Extension Evaluation (Petheram 1998)

- ☆ An evaluation of *need* (system analysis, feasibility study) takes place before programme implementation in order to assist programme design and preparation.
- ☆ Process evaluation focuses on the process of a programme rather than on outcome. Process evaluation is generally carried out during the implementation to assess whether the programme is implemented as planned, whether inputs result in expected outputs through the expected process. Process evaluation is generally formative in nature, but can also be summative, for example when the stakeholders need to know about actual delivery of a programme for justification of programme spending (Petheram 1998). This activity is also called monitoring (Groot and Boon 1992).
- ☆ Outcome evaluation and 'efficiency evaluation' take place both during and after programme implementation. Outcome evaluation is generally summative, but may be formative when findings are used to improve future or other existing programmes (Petheram 1998). Outcome evaluation is also called terminal evaluation. It measures whether expected outcome have been achieved, and assesses to what effects on farmer practice, consequences

for farm management, and impacts these outcomes have led. Evaluation of efficiency considers the relation between programme costs and outcomes (van de Fliert 1993).

3.2.3 Importance of Formative (or Process) Evaluation

The focus on outcomes evaluation (Summative evaluation) is necessary in the context of budget restriction and increasing need to be accountable for public spending. But there are several arguments that highlight the danger of focusing exclusively on summative evaluation. The need for more formative evaluation in agricultural extension underpins the larger movement towards increasing 'process-oriented programmes' in general (Petheram 1998). Programme with a process orientation are difficult to evaluate as they are carried out in variable, unpredictable situations, they produce output that are hard to measure objectively and often have permeable boundaries and less-than-direct relationship between inputs and outputs (Farrington and Nelson 1997).

3.2.4 Formative (Process) Evaluation vs. Summative (Outcome Evaluation)

1. Summative Evaluation

Outcome evaluation (terminal/summative) carried out at the end of a programme may fail to provide information on how the outcomes were achieved, the causal factors for change in attitude difficult to identify. Programmes with strong focus on outcome evaluation, indicators become the mission. The outcome oriented evaluation built into the programme model may drive it into concentrating on producing measurable outcome when perhaps most desirable outcome is not easily measurable (Petheram 1998).

2. Formative Evaluation

The strengths of formative (process) evaluation are: (Petheram 1998)

1. Knowledge gained about the process that led to desired or undesired outcome can be fed into future programme,
2. To identify if programme is not achieving desired outcomes before the programme ends, to give the potential refining, re-defining or clarifying the programme model which requires considerable flexibility in programme,
3. To inform stakeholders what is going on at the ground level, as it is common for plans to change considerably on implementation,
4. To incorporate and cater for unexpected happenings or intermediate outcomes,
5. If done collaboratively, can lead to teambuilding, as it strengthens a common understanding of programme aims, objectives and purpose,
6. It can build in flexibility and iteration into programme models from the onset.

3.2.5 Owen's Five Forms (Types) of Evaluation

Various attempts have been made to classify evaluation *per se*, some by categorising forms of evaluation by purpose (Owen 1993), others by methodology (Stake 1973), and others by the position of the major audience (Worthen *et al.*, 1997). In the context of present study conceptual meta-model of Owen (1993) is discussed to classify the evaluation forms which have been enlisted under section 3.2. Owen's five forms of programmes evaluation are elaborated in Table 3.1. This framework is useful in conceptualising evaluation *per se*.

3.2.5.1 Form 1: Evaluation for Impact Assessment

'Impact evaluation' as evaluation leads to a decision about the worth of a programme which often has strong summative emphasis (Owen 1993). This type of evaluation generally carried out at the end of the programme, or when a programme is at a settled phase. The principle foci of this type of evaluation are:

- ☆ To understand the outcome of the programme,
- ☆ To justify programme spending, and
- ☆ To gain guidance about what to do next.

Within the category of impact evaluation in agricultural extension, the following approaches were found: (Petheram 1998).

- ☆ Goal-based impact evaluation
- ☆ Need-based impact evaluation
- ☆ Comparative economic impact evaluation
- ☆ Impact evaluation for illumination.

Goal or objective-based evaluation judges whether the stated goals or objectives of a programme have been achieved. Tyler (1967) was the first to develop and use goal-based evaluation. According to Owen (1993), the main task in planning a goal-based evaluation is:

- ☆ To determine the 'real' goals of the programme, and
- ☆ To decide how to determine whether the programme has led to the attainment of goals.

The measure techniques in goals-based evaluation must have strong face validity but sometimes it is necessary to use substitute measure or indexes that stand in the place of preferred or ideal measures (Petheram 1998).

Need-based impact evaluation (goal free evaluation) in the strict sense occurs without the evaluator being aware of the stated goals of the programme. Scriven (1967) offered different perspective towards impact evaluation. His argument was that testing the goal of the programme does not tell the worth of that programme for society in general, in that it does not assess the goals themselves. In reality, goal-free evaluation in its strict sense is rarely practiced in programme evaluation, so it can be referred to as 'need-based' evaluation (Petheram 1998).

Table 3.1: Owen's Five Forms of Programme Evaluation

Form or Approach of Evaluations	*Form 1 Evaluation for Impact Assessment*	*Form 2 Evaluation for Programme Management (Monitoring)*	*Form 3 Process Evaluation*	*Form 4 Evaluation for Design Clarification*	*Form 5 Evaluation for Programme Development*
Ultimate Purpose	Justification	Accountability	Improvement	Clarification	Synthesis
State of programme	The programme is settled	The programme is settled	The programme is at developmental stage	The programme is at developmental stage	The programme has not yet been implemented and is at planning stage
Timing of evaluation	After Programme	Can run throughout the programme	During programme implementation	During programme implementation	Before programme is implemented
Focus of evaluation	On outcomes/ delivery	On delivery/ outcomes of the programme	On the delivery of the programme	On the design of the programme	On context of the programme environment
Typical approaches	Objectives-based outcome evaluation, needs-based evaluations, economic assessment of cost-benefit of programme	Programme monitoring, component evaluation, use system evaluation, use of performance indicates over time	Implementation studies, action research, responsive evaluation, topical RRAs or PRAs, process evaluation	Evaluability assessment, programme logic, accreditation	Needs assessment, review of best practice, research, synthesis, exploratory RRA (rapid rural appraisal) or PRA (participatory rural appraisal) ex-anti evaluation
Time span	Programme finish	Throughout programme	Programme is settled	Programme is developing	Programme synthesis

Form 1
Form 2
Form 3
Form 4
Form 5

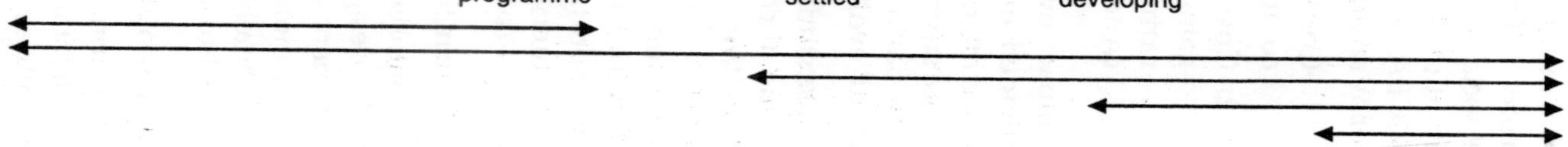

Source: Owen (1993).

Financial analysis of outcome is done by calculating cost-benefit analysis (CBA). It is a procedure for comparing alternative courses of action (or no action) by references to net social benefits that they produce (Petheram 1998). This method is for organising information to aid decision about the allocation of financial resources (Anonymous 1991). CBA can be applied to specific projects or to larger programmes as a whole. Its power as an analytical tool rests on two main features:

1. Costs and benefits are expressed in money terms and hence are directly comparable, and
2. Costs and benefits are valued in terms of the claims they make and the gains they provide to the economy as whole, so the perspective is wider one rather than that of a particular individual, organization or group (Anonymous 1991).

Impact evaluation for illumination aims to explore both the intended and unintended outcomes of programme intervention. Evaluation has a prominent presence in agricultural development in lower income countries (Petheram 1998). There is a strong focus on the use of logical frameworks, and an increasing tendency towards participatory process project design (Farrington and Nelson 1997) and often the word 'evaluation' is replaced by the term 'impact assessment' (Petheram 1998).

Case study evaluation is another technique of impact assessment. Internationally the value of conducting case studies is now recognised by development agencies such as World Bank (Casley and Kumar 1987). The individual is implicit in the notion of 'case', but the individual may be a group, a project, a village, an institution, a region and so on (Sechrest *et al.*, 1996). Case study evaluation is the study of the particularity and complexity of a single case, coming to understanding its activity within important circumstances (Stake 1995). It is thick in description and can reveal how the intervention brings changes in the dependent variable and may be the best design to answer question about an emerging phenomena (Sechrest *et al.*, 1996).

3.2.5.2 Form 2: Monitoring and Evaluation for Programme Management

Monitoring and evaluation (M&E) for programme management (based on Owen's form 2) is designed to run throughout the life of a project and is generally set up at the development stage of the programme. The difference from other forms of evaluation is that it is more discrete (Petheram 1998). M&E for programme managements is conventionally associated with input-output monitoring which would be represented by levels 1 to 3 Bennett's hierarchy (Bennett 1977) (Table 3.2).

3.2.5.3 Form 3: Process Evaluation

The principal focus in process evaluation is on the improvement of a programme and understanding how it is being implemented. It is generally carried out at before programme completion and is distinct from monitoring in that it a discrete activity. Example of this form includes implementation studies, action research, responsive evaluation and topical rapid rural appraisal (RRA) or participatory rural appraisal (PRA) (Petheram 1998). Studies of this kind are referred to as process-outcomes studies and are generally summative in nature (Owen 1993). However in some cases the end point of the process-implementation study is to simply assess the extent of implementation process evaluation carried out to provide guidance in refinement of a programme and has a strong formative element (Petheram 1998).

Table 3.2: Bennett's Hierarchy with Example

Level	*Description*	*Type of Evaluation*
Level 7	Consequences for the target group	Evaluation and Monitoring relating to social/ economic impact
Level 6	Behavioural changes in the target group: (practice changes in the target group indicated by change in quality in products)	Evaluation and monitoring relating to behaviour change (also impact monitoring)
Level 5	Change in knowledge, attitude, motivation and group norms	Evaluation and monitoring relating to intermediate indications of behaviour change
Level 4	The farmer's opinion about extension activities	Evaluation and monitoring relating to farmers' opinion
Level 3	Farmers participation in extension activities (participation monitoring)	Monitoring degree of participation
Level 2	Implementation of the programme by extension agents (activities monitoring)	Activity monitoring
Level 1	Programming of the extension activities (inputs monitoring)	Input monitoring

Modified by Petheram (1998) based on Bennett (1977).

3.2.5.4 Form 4: Evaluation for Design Clarification

It is generally carried out when the programme is at the development stage. Methods used in this form are evaluability assessment (Wholey 1983, Smith 1989). It enables the participants to question the cause and effect assumption between the various steps. It is best used as a team approach and is noted for enabling staff to come to consensus about the realistic outcomes and goals of a programme. Programme logic would be mapped out by project team before the implementation and modified and referred to throughout the life of the project (Petheram 1998).

3.2.5.5 Form 5: Evaluation for Programme Development

This is for the synthesis of a new programme. The programme will not yet have been implemented and will still be at planning stage. The focus is on the context of the programme environment (Petheram 1998). Models and methods include need assessment, economic evaluation, review of best practice, research synthesis, exploratory PRA or RRA and the Delphi technique. Even through Owen (1993) stated that the need assessment is the best known and most acknowledged approach to evaluation for programme development but Scriven (1991) warns that the term 'need assessment' has drifted from literal meaning into Jargon status in which it refers to any study of the needs, wants, market preferences, values or ideas that might be relevant to a programme.

3.2.6 Framework for Programme Evaluation

The ultimate purpose of evaluation which has been termed 'form', but in reality the evaluation of a programme can often employ more than one form. All evaluations have conceptual framework. According to Hulme (1997) there are three main elements to a conceptual framework for an evaluation:

1. The model of the impact chain that the study is to examine,
2. The specification of the unit or levels at which impacts are assessed; and
3. The specification of the types of impact that are to be assessed.

Several logical frameworks are used (currently) in agricultural extension to help plan an evaluation strategy, these are: (Petheram 1998)

1. Bennett's hierarchy: a seven-level ladder that conceptualizes hierarchical levels of programme outcome (Table 3.2),
2. The logical framework a hierarchy of objective statements regarding the goal, purpose, output, and inputs (Table 3.3),
3. The Snyder model- a soft system approach to evaluation (Figure 3.2), and
4. The three-rings approach-differs from the other frameworks in that it recognizes that other agencies and programmes may also have an impact on the target group (Figure 3,3).

3.3 Methodologies for Evaluation of IPM Programmes

IPM is one of the most widely used catchwords in agricultural development and environment conservation programmes. Everybody claims to like IPM and even to do IPM, but the actual content of this term differs widely with the added concern that chemical companies redefine the term IPM in order to boost pesticide sales (Waibel *et al.*, 1999). Waibel *et al.* (1999) observed that the vast range of IPM interventions exist, which make it necessary to improve the understanding of the true impact that can be expected from IPM and hence, the question of measuring the success of IPM programmes is crucial. Workshop on 'Evaluation of IPM Programmes Concepts and Methodologies' was held at Hannouver, March 16-18, 1998 in cooperation with FAO. There were four major conclusions drawn at the workshop. These were:

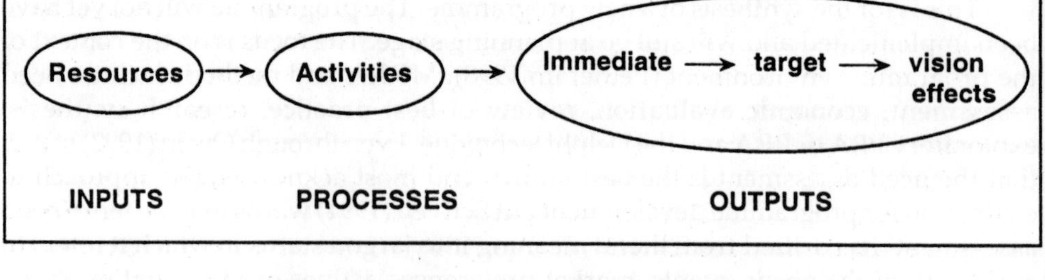

Figure 3.2: Synder Approach to Evaluation (cited from Petheram 1998)

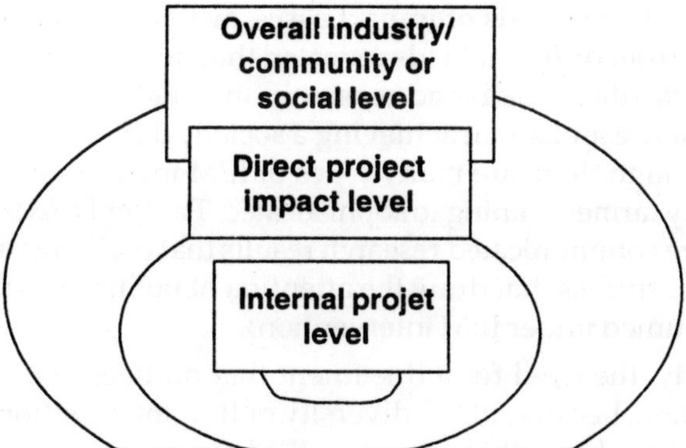

Figure 3.3: The Three Things of Extension Evaluation (Coutts 1997)

Table 3.3: The Log Frame Matrix: Questions to be Answered when Filling in Each Cell of a Log Frame

	Narrative Summary	*Measurable Indicators*	*Means of Verification*	*Assumptions*
Broader goal	What wider issue will the project help to resolve (*e.g.* a national objective)?	What are the indicators of goal achievement?	What are the sources of information? What methods are to be used to obtain it?	**(Purpose to goal)** What assumption/ external factors must be true if the purposes are to help reach the goal?
Purpose [Primary reasons for the project]	What immediate changes/benefits will the project bring about for the target population or area?	What are the measures to judge the project's immediate effects, benefits and losses?	What are the sources of information? What methods are to be used for obtaining it?	**(Outputs to purpose)** What external factors must be true if outputs are to achieve the purpose?
Outputs	What direct outputs will the project produce so it can achieve its purpose?	What kind and quantity of outputs are planned, and by when will they be produced?	What are the sources of information? What methods are to be used for obtaining it?	**(Inputs to Outputs)** What external factors must be realised to produce the planned outputs in time?
Activities	What activities must be under-taken to produce the outputs? When must these activities take place?	**Inputs/resources** What materials equipment or services are required, at what cost, over what period?	What are the sources of information on inputs? What methods are to be used to obtain it?	**(Initial assumptions)** What actions outside the control of the donor are necessary to begin the project?

Source: Farrington and Nelson (1997).

- ✰ "Firstly, the methods of impact assessment and evaluation presented by experts from different fields revealed that despite for the many definitions about IPM, there was broad consensus that IPM is about avoiding pesticides dependence and about achieving a socially defined level of pesticides use even though there are many types of IPM intervention. They range from ordinary farmer training to sophisticated 'Farmer Field School' approaches. Effective communicated research results that challenge existing paradigms and case studies that draw the attention of politicians and civil society can be subsumed under IPM intervention.
- ✰ Secondly, the need for a document that outlines good practices for IPM evaluation, because of the diversity of IPM intervention that already exist and that can be further expected. IPM can only be measured by its results and no longer by its inputs as virtually all groups involved in crop protection, including the chemical companies, label their concept IPM.
- ✰ Thirdly, the different groups who participated in the workshop advocated different approaches to evaluation. There were those who saw cost benefit analysis as the major tool of evaluation and wanted to refine this tool by adding economic evaluation of environmental impacts and by applying contigent valuation methods and willingness to pay concepts. Others proposed that evaluation should not be checking procedure but should be designed in the sprit of joint learning experience. There was significant difference between economists on the one hand and anthropologists and extension specialists on the other hand. There is however, a need to set standards which must be met if cost benefit analysis is going to provide the information that facilitates decision making.
- ✰ Finally, there is a problem of measurements. Hence, unlike some of the 'green revolution technologies' assessing IPM impact is not a matter of adoption versus non-adoption. It is a continuum that is expected to induce a number of processes which may not always be easy to measure because of time lag with which they may occur and because of lack of indicators. Thus inferences drawn from measurable short term income effects induced by IPM may be wrong if evaluation fails to capture processes that can lead to much higher income effects in the future. Therefore additional methods of evaluation can reduce the uncertainty inherent in the assumption entering cost benefit analysis and can provide additional criteria for assessment. These methods of IPM community, *e.g.* cost benefits analysis experts often go beyond form surveys with standard questionnaires and are based on participant observation and case studies" (Waibel *et al.*, 1999).

The guidelines on good practice of IPM evaluation and impact assessment require several coordinated steps. The different groups, plant protection people, and extension specialists have been using different sets of methodologies to measure impact, thus results were not comparable and had little impact on actual decision making (Waibel *et al.*, 1999).

3.3.1 Economic Impact Analysis

Cost-benefit analysis is well known method that has been used on its own or with other methods across a number of forms of evaluation, but is not a form in itself. It is the approach within the evaluation for impact assessment (Form 1 of Owen's meta model; Owen 1993).

IPM economic impact can be evaluated at two levels: the basic farm level analysis and at programme level (Feder and Quizon 1999). Farm level analysis investigates whether IPM and its dissemination-change farmer's knowledge and thereby affect more efficient pesticide use to result in higher farm profit. At programme level, analysis shall focus on overall benefits of state funded IPM diffusion efforts and compare these with overall cost. For IPM and its dissemination, the desired impact on profits comes from raising farmer's knowledge and this rise in knowledge leads to change in the input mix and practice used; and in particular, to a small use of pesticides (Feder and Quizon 1999). Feder and Quizon considered assessing the farm level impact of IPM and its diffusion a complex task for which before-and-after or with-and-without comparative study may not suffice and recommended, two stage and simultaneous equation procedures to correct for endogeneity and self-selection problem.

3.3.2 Goal-based Impact Assessment of IPM

IPM is a knowledge intensive technology, it needs more suitable approach than that which has been commonly been applied in adoption studies (Chung *et al.*, 1999). IPM is not a single decision rule, but interactive inter-linked concepts which can not be measured, for-example, a binary variable (Adoption/non-adoption). It is dynamic continuum, implying a more complex continuum, implying a more complex relationship between knowledge acquisition and farmer practice and difference in crop production decisions may arise from different levels of IPM competency, for this structural questionnaires and analytical techniques to derive indices of IPM knowledge that may vary both within and among groups of IPM and non-IPM farmers (Chung *et al.*, 1999).

3.3.3 Participatory Evaluation of IPM Programmes

Waibel (1999) discussing the methodology of IPM evaluation argued that the evaluation needs a policy perspective if the true impact of IPM is to be captured. The success of IPM programmes/projects must be measured relative to their objectives. These depend on:

1. A country, a region or a cropping system is already dependent on pesticides with strong indicators of overuse and misuse. The major goal of IPM must be to reduce pesticide use to socially optimal level and get the system out of this dependency, and
2. In area or agricultural system with zero or low levels of pesticides use, the aim of IPM project will be to avoid the system to get into pesticide 'treadmill'.

According to Waibel (1999) on the farmer level indicators of economic benefits of IPM programme are:

- ☆ Saving on pesticides

- ☆ Increased yield
- ☆ More stable income
- ☆ Increased business opportunities
- ☆ Improved health status

The framework for participatory evaluation of IPM given by Waibel (1999) also emphasizes that measuring the success of IPM must accommodate its dynamic nature. A successful IPM intervention will induce a process that not only leads to better crop management, decision making but also stimulates a discovery process, strengthens the build-up of institutional capacities at village level and intensifies policy interaction (Figure 3.4).

3.3.4 Evaluation of Impact of IPM Programmes at the Farm Level

The IPM evaluation at the farm level covers wide variety of sub-topics. The possible areas were identified as subject to evaluation by Working Group I for "Evaluation of IPM programmes: Concepts and Methodologies" at Hannover, March 16-18, 1998 (Waibel *et al.*, 1999).

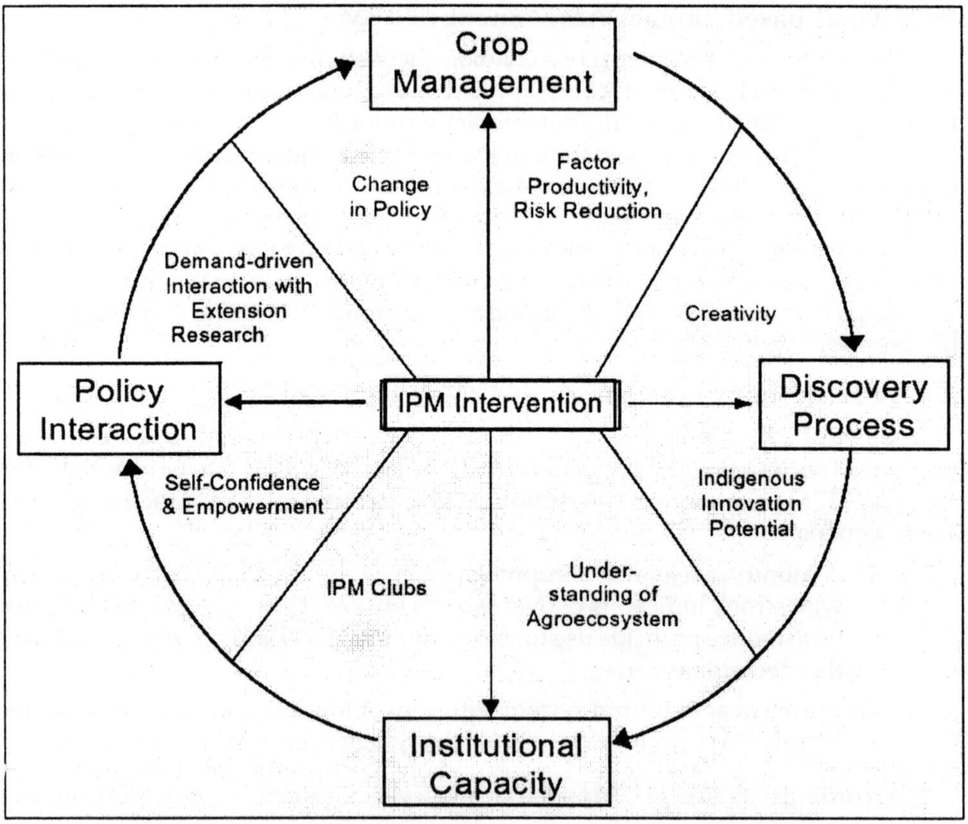

Figure 3.4: Framework for Participatory Evaluation of IPM Projects
***Source*: Waibel *et al.* (1999)**

The subjects of evaluation are:

1. Improved of economic well-being;
2. Adoption of new technologies (passive); adaptation of technologies to local conditions (active);
3. Improved knowledge and analytical capacity;
4. Diffusion of knowledge farmer to farmer;
5. Decreased health risk, and
6. Healthier ecosystem.

Requisite methods vary enormously depending on the goals of the evaluation and the resources available for the study.

Table 3.4: Data and Methods Required for the Evaluation of IPM Programme at Farm Level (Priority Issues)

Sl.No.	*Results*	*Methods of Data Collection (Before/After, With/Without Project)*
1.	Adoption of new technologies (passive); adaptation of technologies to local condition (active)	☆ Short questionnaire, part of household survey ☆ Controlled comparison have two or more geographical area with different situations ☆ Direct observation of pesticide containers (amount, kind, storage, disposal) ☆ Participant observation is the best method for teasing out complex relations eg. behaviour change ☆ Interviews ☆ Drawings ☆ Direct observation of adaptation ☆ Interviews with key informants ☆ PRA-type methods with extensionists (first 5 results) ☆ Survey of local pesticide sellers (volumes and types sold) ☆ National pesticide statistics
2.	Improved knowledge and analytical capacity	☆ Before and after knowledge test of farmers ☆ Correlation of knowledge and practice
3.	Improved economic well-being	☆ Farm budget survey on income and expenditure

Source: Waibel *et al.*, 1999.

The formal evaluation exercises (formative as well as summative) needs to be carried out within the context of IPM projects. Three formal socio-economic surveys within the evaluation period are recommended: first a baseline survey before the project starts second survey after the first year and third survey (final survey) after completion of project which includes the overall data analysis.

Waibel *et al.* (1999) also reported about the recommendations of working groups II, III and IV about the evaluation of IPM projects at village level, impacts on institution and policies, impact on natural resources and environment, respectively. The last two: impact on institution and polices and impact on natural resources and environment are not under the scope of this research study. The task force meeting on assessing the impact of participatory research and extension methods in IPM was also held at Nairobi; 15-17 March, 2001 (Ortiz 2001).

The report of working group II regarding phases and indicators for evaluations of impact assessment are given in Table 3.5. The report concluded that evaluating the impact of agricultural technology is a challenge, and evaluating the impact of an of information-knowledge intensive technology such as IPM is even more challenging (Ortiz 2001).

Table 3.5: Phases and Indicators for Impact Assessment

Phase	*Needs and Opportunity Assessment*	*Technological Options*	*Learning Model for, and Integration, Adaptation of Options*	*Implementation of Protocols (Pilot Scale)*
Type of assessment	☆ PRA ☆ Baseline ☆ GIS ☆ Modeling	☆ Technology evaluation modeling	☆ Technology evaluation ☆ Process evaluation ☆ Impact evaluation ☆ Modeling	☆ Process evaluation ☆ Impact evaluation ☆ GIS ☆ Modeling
Indicators				
Natural	☆ Yield ☆ Pest levels ☆ NE activity ☆ Pesticide load ☆ Soil health	☆ Yield ☆ Pest levels ☆ NE activity ☆ Indicators of performance	☆ Yield ☆ Pest levels ☆ NE activity	☆ Yield ☆ Pest levels ☆ NE activity ☆ Pesticide load ☆ Soil health
Human	☆ Participation ☆ IPM knowledge ☆ IPM skills ☆ IPM practice ☆ Health ☆ Gender ☆ Poverty	☆ Participation ☆ Farmer perception	☆ Participation ☆ IPM knowledge ☆ IPM skills ☆ IPM practice	☆ Participation ☆ IPM knowledge ☆ IPM skills ☆ IPM practice (adaptation, adoption, rejection) ☆ Health ☆ Gender ☆ Poverty
Social	☆ Collectivity ☆ Awareness ☆ Regulatory framework	☆ Collectivity	☆ Collectivity ☆ Dissemination ☆ Awareness	☆ Collectivity ☆ Dissemination ☆ Awareness ☆ Regulatory framework
Financial/ physical	☆ Farm economic analysis ☆ Supporting infrastructure	☆ Farm economic analysis	☆ Farm economic analysis	☆ Farm economic analysis ☆ Supporting infrastructure ☆ Cost-benefits analysis

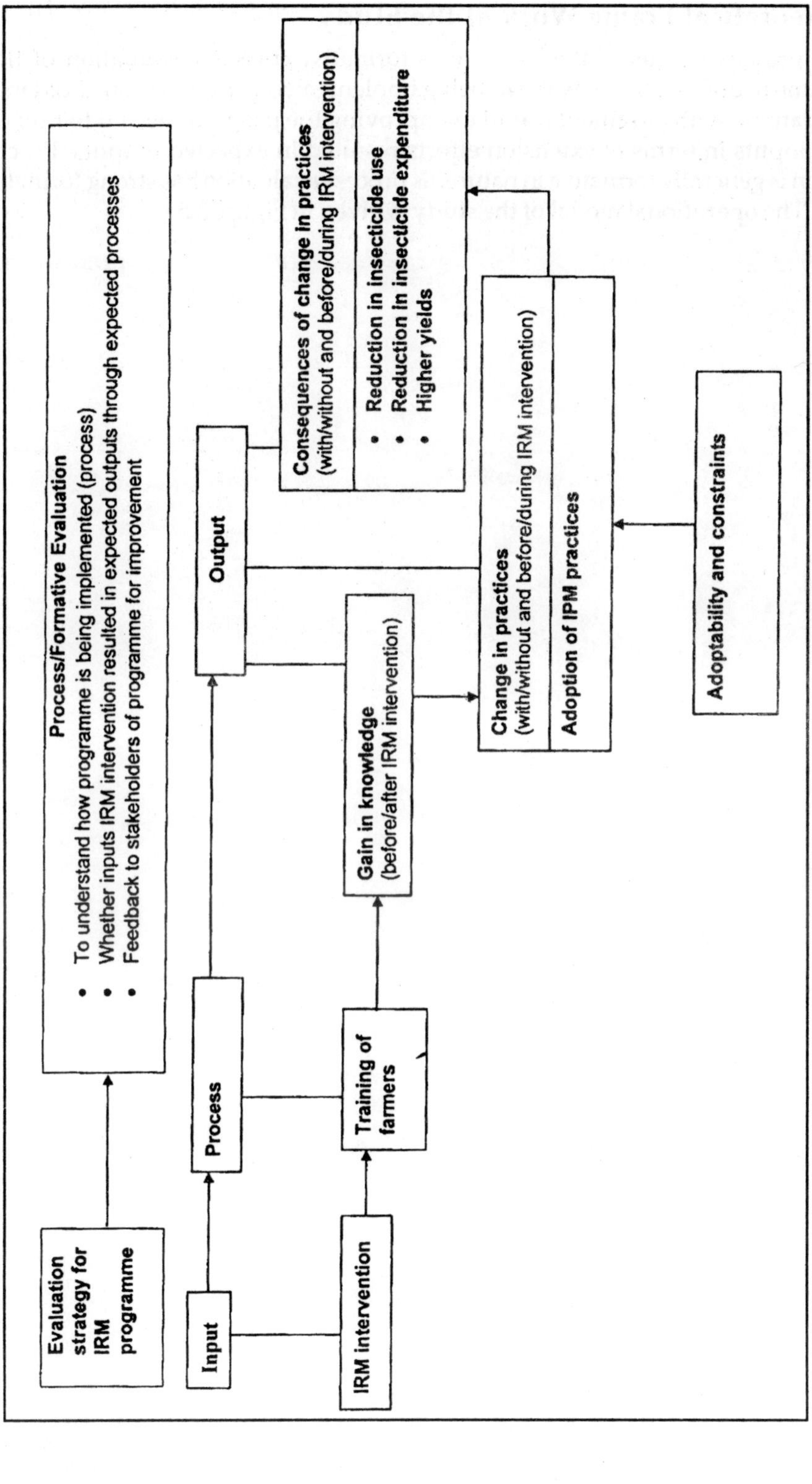

Figure 3.5: Operational Model of the Study

3.4 Theoretical Frame Work of the Study

The principal focus of this study was formative/process evaluation of IRM programme to understand how it was being implemented to provide stakeholders of the programme with judgment useful in improving the programme and to find out whether inputs in terms of extension efforts resulted in expected outputs. Process evaluation is generally formative in nature, as process evaluation has strong formative element. The operational model of the study is given in Figure 3.5.

Chapter 4
METHODOLOGY FOR FORMATIVE EVALUATION

This chapter deals with the various aspects of the research methodology followed in conducting the study. The research design, methods and procedures followed for conducting the study are discussed here under:

4.1 Research Design of the Study

Quasi-experimental cum survey design of research was employed for conducting the formative/process evaluation of the Insecticide Resistance Management programme. With and without and before and during experimental design was applied in villages covered under IRM programme (experimental group–with IRM intervention) and villages not covered under IRM programme (control group–without IRM intervention) for studying the effect of IRM programme. However, there was an error in not applying with and without design for measurement of knowledge level, and knowledge gain was measured by applying pretest (before) and post-test (after) in case of experimental group. This error has resulted in employing faulty experimental design. Experimental group and control group here onwards will be referred to as IRM villages and non-IRM villages, respectively.

4.2 Locale of the Study

The study was conducted in the state of Punjab. Three cotton growing districts namely Bathinda, Ferozepur and Mansa were purposively selected for the study as these three districts are being covered under the IRM project.

4.3 Sampling Plan of the Study

In Punjab, training of farmers under IRM programme was first started in 2002. In 2002, 30 villages (ten from each district) and in 2003, 30 villages (ten from each district) were covered under IRM programme. In 2004, 45 villages (15 from each district) were taken up for disseminating the technology under IRM project. Five villages from each district were randomly selected out of the villages covered under IRM programme 2004. Thus, a total sample of 15 IRM villages was selected for the study. The IRM villages selected were Bagwangarh, Gurusar Sainewala, Malwala, Machhana, Mehta (district Bathinda); Bahwal Wasi, Bhangala, Goddan Doab, Rajpura, Wahabwala (district Ferozepur); Bhaman Khurd, Dasahndia, Gharangne, Khaiali Chahlanwali and Maujian (district Mansa). These 15 IRM villages formed the experimental group.

Out of the nine non-IRM (control) villages selected under the project (three from each district), two villages from each district were selected randomly, and thus six non-IRM villages were taken up for the study. The non-IRM villages selected for the study as control group were Deepa Bangi, Sangat Kalan (district Bathinda); Doda, Kundal (district Ferozepur); Gagowala and Nangal Khurd (district Mansa). Thus, 21 villages (15 IRM, 6 non-IRM) in all were selected for the study. The six non-IRM villages selected for the study formed the control group. Map showing the study villages is given in Figure 4.1.

4.4 Selection of the Respondents

Under IRM programme, ten cotton growers from each village were forming the core group for training and monitoring the insecticide resistance in the field. All these ten farmers from each of the selected IRM villages were selected purposively for the study. The total sample of cotton growers selected from 15 IRM villages was 150, who here onward will be referred to as IRM farmers.

The list of cotton growers compiled by IRM project staff in six selected non-IRM villages was Deepa Bangi–34, Sangat Kalan–60, Doda–64, Kundal–86, Gagowala–68 and Nangal Khurd–60.

From these six non-IRM villages, ten farmers from each village were selected randomly and the total sample size of farmers selected from these non-IRM villages was 60, who here onwards will be referred to as non-IRM farmers. Thus the total sample size of farmers was 210 (150 IRM farmers and 60 non-IRM farmers).

For studying knowledge gain pretest-post-test experimental design was followed only in IRM villages. To find out the adoption and economic benefits of IPM practices disseminated under IRM project, the comparison was made between IRM village (experimental group) and non-IRM villages (control group). The adoptability and the constraints of IPM practices disseminated under IRM were studied only in IRM villages.

4.5 Selection of IPM Practices under IRM Programme

The IPM practices enlisted in the IRM strategy, under the programme were selected for the study. IPM practices selected were cultural practices, chemical control practices and conservation of natural enemies as discussed in chapter 1, sections 1.2.1, and 1.2.2 and1.2.3.

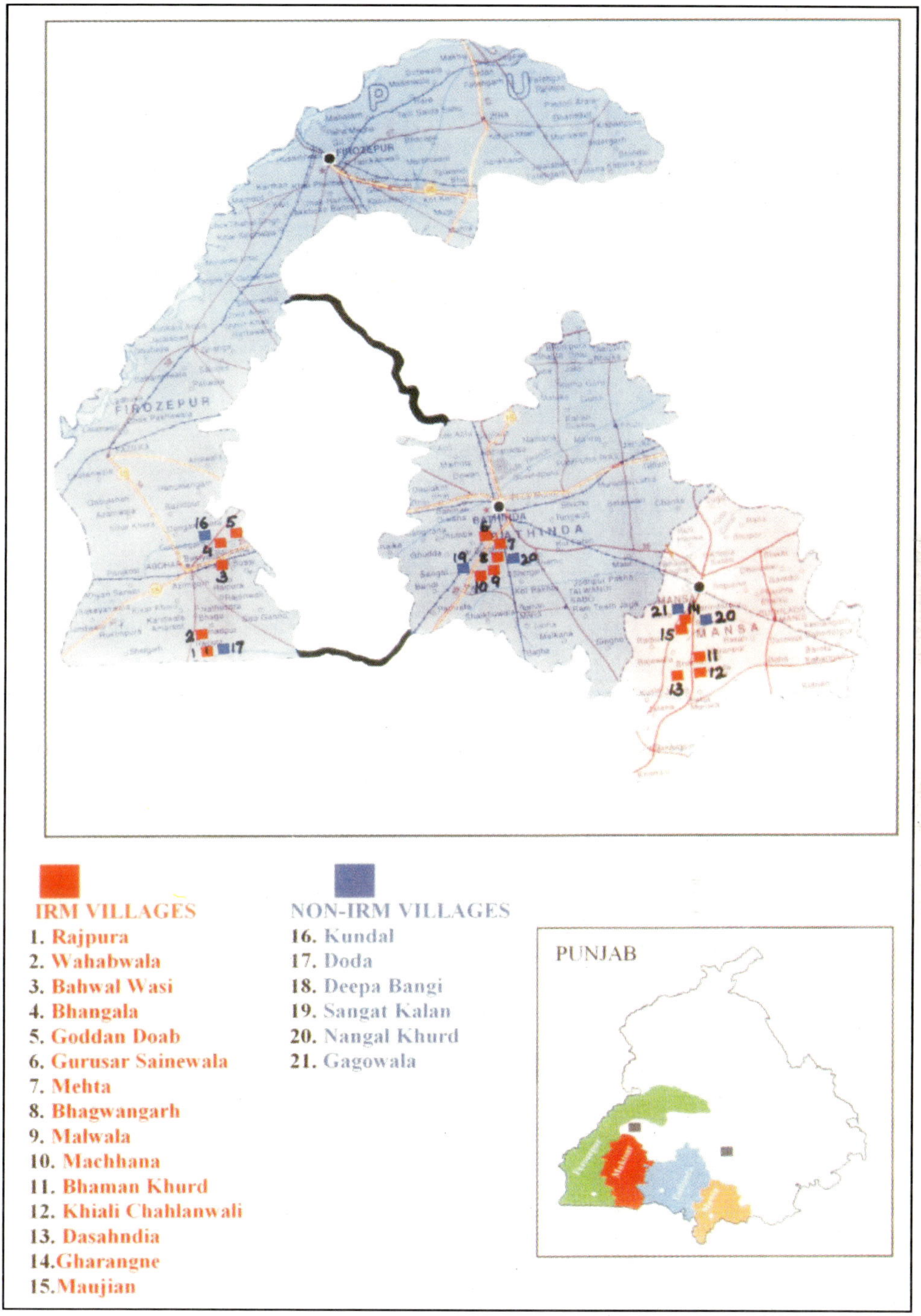

Figure 4.1: Location of Study Villages in Punjab

4.6 Operational Definitions

There are in general, two types of operational definitions: measured and experimental. A measured operational definition describes how a variable will be measured and an experimental operational definition spells out the details of researches manipulation of a variable (Kerlinger 1973). Operational definitions related to experiment and measurement is given under:

4.6.1 Experimental Group

IRM villages were selected as experimental group for finding out the effect of IRM programme (with), in terms of adoption of different IPM practices, reduction in the use of insecticides (technical grade material), insecticide applications, insecticide expenditure and yield. It was measured by employing with and without, and before and during research design. The effect of IRM intervention in terms of gain in knowledge was measured in experimental group using before and after design.

4.6.2 Control Group

Non-IRM villages acted as control group (without IRM programme) for finding out the effect of IRM programme in IRM villages. It was measured in terms of difference in IPM practices adopted, insecticide use (technical grade material), insecticide applications, insecticide expenditure and yield between control group and experimental group.

4.6.3 Knowledge Gain About Selected IPM Practices

It is the change in the number of correct responses given by IRM farmers before and after (pre-test and post-test) IRM training programme by way of recall or recognition or identification on 29-item standardized knowledge test constructed for the study.

4.6.4 Training Strategy

It has been evaluated in terms of the extension methodology followed (instructional process) for dissemination of selected IPM practices with respect to trainers, materials, training aids and overall effectiveness.

4.6.5 Extent of Adoption

It is a measure of adoption of selected IPM practices dissemination under IRM strategy, whose adoption has been measured in terms of percentage of area covered and percentage of farmers adopting a particular practice (Singh and Peshin 2000)

4.6.6 Level of Adoption

It is with respect to adoption of those selected IPM practices which were divisible and were measured in terms of deviation from the recommended quantities of inputs *viz.* dosages, seed rate (Singh and Peshin 2000).

4.6.7 Economic Benefit

It is a measure to study the economic benefits of adoption of selected IPM practices and was measured with respect to reduction in terms of variable cost *viz.* pesticide used and increase in the yield per unit area by IRM farmers in comparison to non-IRM farmers.

4.6.8 Adoptability

It is with respect to suitability of selected IPM practices for adoption and was measured with respect to different attributes of selected IPM practices by way of responses of IRM farmers to open ended questions and on a three point continuum scale having scoring scheme as indicated below:

Completely:	2
Partially:	1
No:	0

Responses by way of can not say/do not know were not scored for calculating the index of an attribute.

$$\text{Index of an Attribute} = \frac{\text{Maximum score obtained}}{\text{Maximum score obtainable}} \qquad (i)$$

It can range between 0 to 1

Adoptability index was worked out by:

$$\text{Adoptability index} = \frac{\sum \begin{array}{l}\text{Indices of positively related}\\ \text{attributes of an innovation}\end{array} - \sum \begin{array}{l}\text{Indices of negatively related}\\ \text{attributes of an innovation}\end{array}}{\text{Number of attributes}} \qquad (ii)$$

It can range between–1 to +1.

4.6.9 Constraints

It is a measure of limitations faced by the farmers in the adoption of selected IPM practices and was measured in terms of percentage of farmers reporting the constraint in the adoption of a particular practice.

4.6.10 Education

It was measured in terms of number of years of formal education of the farmers and categorized into illiterate, up to primary, middle, matriculate, 10+2, graduate and above.

4.6.11 Operational Land Holding

It refers to the number of hectares owned by the farmers, including leased in but excluding leased out.

Operational land holding = Area owned + Area leased in–Area leased out

The operational land holding of farmers were categorized into:

Small:	1-2 ha
Semi medium:	2-4 ha

Medium: 4-10 ha

Large: > 10 ha

The categorization of land holding is based on the categories used by Anonymous (1987, 1994) and Johl (1997).

4.7 Construction of Research Schedules for Data Collection

Three research schedules were constructed for collection of the data on different aspects of the study.

Research schedule number one consisted of two parts, part I: profile of the respondents and base line data of 2003, part II: knowledge test (Table 4.1). Research schedule number two was in the form of semi-structured questionnaire for studying the adoption of different practices (Table 4.2). Research schedule number three consisted of two-parts. Part I was to study adoptability and constraints in the adoption of selected IPM practices, and Part II was for evaluation of the IRM training programme in terms of training strategy (Table 4.3).

4.7.1 Construction of the Knowledge Test

For construction of the knowledge test for the farmers about selected IPM practices the pattern followed by Bhatia (1990), and Peshin and Kalra (2000b) for developing a knowledge test was followed, who had followed the pattern of using open-ended, multiple choice, along with questions for recognition/identification of specimen and objects.

4.7.2 Procedure for Construction of Knowledge Test

Construction of knowledge test consisted of the following steps:

4.7.2.1 Definition of knowledge,

4.7.2.2 Statement of process objectives,

4.7.2.3 Content area, and

4.7.2.4 Preparation of tests blue print and writing test items.

4.7.2.1 Definition of Knowledge

The knowledge for this purpose was defined as "Behavior and test situations which emphasize the remembering, either by recognition or recall, of ideas, materials or phenomena" (Bloom 1979).

4.7.2.2 Statement of Process Objectives

The process objectives selected were in terms of the following cognitive processes of the farmer in the field of IPM technology disseminated under IRM strategy.

1. Recalls or recognizes the terms, concepts, specimen (knowledge of specific).
2. Knowledge of ways and means of dealing with specific problems.
3. Knowledge of the universals and abstractions in a particular field.

Table 4.1: Research Schedule 1- Knowledge Test

Department of Extension Education
Punjab Agricultural University Ludhiana

Title of the Research Problems: "Evaluation of Dissemination of Insecticide Resistance Management Technology in Cotton Crop in Punjab".

Sl.No.: ____________________

Village: ____________________

Block: ____________________

Date: ____________________

PART I: Profile of the Respondent and Base Line Data

1. Farmer's Name: ______________________________
2. Father's Name: ______________________________
3. Telephone number (if any) ______________________________
4. Education
   (*a*) Illiterate
   (*b*) Can read
   (*c*) Can read and write
   (*d*) Primary
   (*e*) Middle
   (*f*) Matric
   (*g*) 10+2
   (*h*) Graduate and above
5. Operational Land Holding (in acres)

	Irrigated	*Unirrigated*	*Total*
(*a*) Owned	________	________	________
(*b*) Leased in	________	________	________
(*c*) Leased out	________	________	________
Total operational Land hoding $(a + b) - c$	________	________	________

	Baseline Data	*During 2003*	*During 2004*
6.	Area under cotton crop (in acre):	________	________
7.	Annual family incomes Rs.	________	________
8.	Number of pesticide sprays applied in cotton crop in 2003	________	________
9.	Expenditure on pesticide sprays/acre during 2003 Rs.	________	________ (Including labour)
10.	Average yield of cotton/acre during 2003	________	________ (in kg)
11.	Sale rate of cotton/quintal during 2003 Rs.	________	________

Contd...

Table 4.1–Contd...

12. Varieties of cotton sown during 2004: 1)________, 2)________
3)________, 4)________
5)________, 6)________

13. Sowing of cotton crop during 2004: completed by which date? ____________________

PART II: Knowledge Test

Weightage	*Test Item*	*Before IRM Project*	*After IRM Project*
(2)	1. Name two recommended resistant varieties of each of:	(i) American Cotton: ________ ________ (ii) Hybrid Varieties: ________ ________	 ________ ________ ________ ________
(1)	2. Sowing of cotton should be	*April/May* ________ ________	*April/May* ________ ________
(1)	3. What is the benefit of timely sowing of cotton crop?	________ ________	________ ________
(2)	4. What should be R x R spacing PxP spacing in case of hybrids?	________ ________	________ ________
(1)	5. How many days after sowing thinning should be done to maintain plant spacing	(i) 30 days (ii) 45 days (iii) 60 days	(i) 30 days (ii) 45 days (iii) 60 days
(2)	6. Recommended fertilizer and manure schedule per acre of cotton crop (in kgs./bags)	*Urea DAP/SSP MOP FYM* i) Hybrid ________ ii) Other varieties________	*Urea DAP/SSP MOP FYM* i) Hybrid ________ ii) Other varieties________
(2)	7. Which are the alternate host weeds for spotted bollworm?	(i)________ (ii)________	(i)________ (ii)________
(1)	8. Early maturing varieties escape the excessive damage from which two insect-pests ?	(i)________ (ii)________	(i)________ (ii)________
(1)	9. Cleaning of bunds/water channels of weeds should be repeated after.	(i) 30 days (ii) 45 days (iii) 60 days	(i) 30 days (ii) 45 days (iii) 60 days
(2)	10. Removing weeds reduce the initial build up of which pests ?	(i)________ (ii)________	(i)________ (ii)________
(2)	11. What is the seed rate per acre ?	(i) American Cotton Kg/acre (ii) Hybrid Cotton Kg/acre	(i) American Cotton Kg/acre (ii) Hybrid Cotton Kg/acre

Contd...

Table 4.1–Contd...

Weightage		*Before IRM Project*	*After IRM Project*
(1)	12. Name the insect-pest whose population builds up on account of excessive irrigation?	________ ________	________ ________
(1)	13. Name the seed treating chemicals which reduce the incidence of diseases ?	(i)________ (ii)________	(i)________ (ii)________
(1)	14. Against which insect-pest seed treatment with Gancho 70WS @ 5gm/kg of seed helps in delaying the first spray up to 60 days.	________ ________	________ ________
(1)	15. How many days after sowing should the first spray be given if required ?	(i) 45 days (ii) 60 days (iii) 90 days	(i) 45 days (ii) 60 days (iii) 90 days
(3)	16. In case of need which pesticide should be sprayed for ABW before 90 days after sowing?	________ ________	________ ________
(1)	17. What is the ETL of the following pests ?	(i) Jassids ________ (ii) White fly ________ (iii)ABW ________	(i) Jassids ________ (ii) White fly ________ (iii)ABW ________
(1)	18. How can you find the ETL for ABW	________ ________	________ ________
(1)	19. Over use of synthetic pyrethroids lead to excessive build up of.	White fly/ABW	White fly/ABW
(1)	20. Which insecticide should be used if whitefly becomes a serious problem during fruiting phase?	________ ________	________ ________
(11)	21. Identify the pest/natural enemies from the specimen:	Whether pest or natural enemy	Whether pest or natural enemy
	(i) Jassid	________	________
	(ii) Whitefly	________	________
	(iii) Curled lead photograph of Jassid	________	________
	(iv) American bollworm	________	________
	(v) Spotted bollworm	________	________
	(vi) Spider	________	________
	(vii) Coccinellids	________	________
	(viii) Predatory bugs	________	________
	(ix) Chrysopa species	________	________
	(x) Cotton aphid	________	________
	(xi) Tobacco caterpillar	________	________
(1)	22. Should pesticides be tank mixed ?	Yes/No	Yes/No
(1)	23. Should group of pesticides be rotated?	Yes/No	Yes/No

Contd...

Table 4.1–Contd...

Weightage		*Before IRM Project*	*After IRM Project*
(3)	24. If yes, name the three pesticides belonging to different groups?	(i)____ (ii)____ (iii)____	(i)____ (ii)____ (iii)____
(2)	25. What are the reasons for reduction in efficacy of insecticides?	(i)____ (ii)____	(i)____ (ii)____
(1)	26. Name the insecticide which is not effective against American bollworm?	(i)____	(i)____
(3)	27. Name the insecticide which can be used against:		
	(i) Young larva of ABW	____	____
	(ii) Old larva of ABW	____	____
	(iii) Incase of outbreak of ABW	____	____
(1)	28. Is spray technique important in pest management?	Yes/No	Yes/No
(1)	29. Are you aware of insecticide resistance?	Yes/No	Yes/No

Table 4.2: Research Schedule 2–Questionnaire

Department of Extension Education
Punjab Agricultural University Ludhiana–141004

(Translated version of the questionnaire)

Part–1: Personal Information

1. Name of the Farmer: ____________________
2. Name of the Father:____________________
3. Village ____________ Block ____________
4. District ____________Telephone no.____________
5. Cultivable land (acre)

   Land Owned ____________________

   Leased In ____________________

   Leased Out ____________________
6. Area under cotton during year 2004 ____________acre

Part–2: Adoption of different recommended practices and inputs use in American and Desi Cotton cultivation

Preparation of Land preparation and Sowing

1. Please, note down the dates of harvesting of wheat crop.

   (i) Date of start of harvesting of wheat ____________

   (ii) Date of completion of harvesting wheat crop____________

Contd...

Table 4.2–Contd...

2. Please, note down the information regarding the use of FYM in American/*Desi* cotton.
   (i) Total quantity of FYM applied ____________ trolleys per acre.
   (ii) Area covered ______________ acre
   (iii) Cost of one trolley of FYM Rs: ______________
   (iv) Size of the trolley.
      Length __________feet
      Breadth __________feet
      Height __________feet
3. Please note down the information regarding different land preparation operation for sowing of American/*Desi* cotton.
   (i) Total number of ploughings _____ and expenditure __ (in Rs./acre)
   (ii) Total number of plankings _____and expenditure _____ (in Rs./acre)
   (ii) Cost of pre-sowing irrigation __________ (in Rs./acre)
4. Total expenditure incurred on preparation of one acre of land Rs ________ (Including hired and family labour).
   Note: Please include the expenditure of family labour in labour expenditure.
5. i. Please note down information regarding sowing of American/Desi cotton.

Sl.No.	*Date of Sowing*	*Method of Sowing*	*Area Sown (acre)*	*Varieties Sown*	*Quantity of Seed per Acre (kg)*	*Cost of Seed per Acre (Rs.)*	*Source of Seed*
1.							
2.							
3.							
4.							

   (ii) Please note down the number of labourers requires for sowing one acre of land ____________

1. (i) Did you treat the seed with seed treating chemicals before sowing? yes/no (tick the ☑ correct one)

   If yes, please note down the following information.

Sl.No.	*Name of Seed Treating Chemicals*	*Quantity of Seed Treated (kg)*	*Quantity of Seed Treating Chemicals Used (g/kg)*	*Total Cost of Seed Treating Chemicals (Rs.)*	*Labour Expenditure for Seed Treatment (Rs.)*
1.	Emison + Stretocycline				
2.	Gaucho				
3.	Any Other i. ii.				

Contd...

Table 4.2–Contd...

(iii) If not, was the seed already treated? Yes/No

(Tick the ☑ correct one)

2. Please note down the spacing maintained while sowing American/*Desi* cotton.

Sl.No.	*Spacing*	*Non Hybrid Varieties of American Cotton*	*Hybrid Varieties of American Cotton*	*Desi Cotton*	*Bt Cotton*
1.	Row to Row				
2.	Plant to Plant				

Fertilizer Usage

1. Please note down the information regarding fertilizers used in American, *Desi* and *Bt* cotton as per detail given below.

Sl.No	*Varieties*	*Urea (Bag/kg per acre)*	*DAP/SSP (Bag/kg per acre)*	*MOP (Bag/kg per acre)*	*Zinc Sulphate (kg)*	*Area Covered*
1.	Non-hybrid varieties of American Cotton					
2.	Hybrid varieties American Cotton					
3.	*Desi* Cotton					
4.	*Bt* cotton					

2. (i) Did you apply urea in split doses ? Yes/No

(Please tick ☑ the correct one)

(ii) If yes, please note down the following information.

Sl.No.	*No. of Split Doses*	*Quantity of Urea Applied (kg)*
(i)	One	________
(ii)	Two	________
(iii)	Three	________

3. Please note down the expenditure incurred on applying fertilizer in one acre ________ (Rs.)

Weed Management

1. Please give the information regarding method of weed control in water channels, bunds and field. (Please tick ☑ the correct one)

(i) Manually

(ii) Mechanically

(iii) Chemically

Contd...

Table 4.2–Contd...

2. (i) Number of hoeing done for control of weeds in American/*Desi* cotton ____
   (ii) Method of hoeing (please tick ☑ the correct one)
   - ☆ Manually
   - ☆ Tractroized
   - ☆ Both

   (iii) Expenditure incurred for hoeing for one acre of land manually Rs. ____
   (iv) Expenditure incurred for hoeing one acre of land with tractor Rs. ____
3. If you use herbicides for weed control, then please note down the following information.

Sl.No.	*Name of Herbicides*	*Quantity Used per acre*	*Total cost of Herbicides (Rs)*
1.			
2.			
3.			

4. Cost of herbicides application ________________ (Rs per acre)

Irrigation

1. Source of irrigation (Please tick ☑ the correct one)
   (i) Tubewell
   (ii) Canal water
   (iii) Tubewell and canal water (both)
2. Total number of irrigation given to cotton
   (i) Hybrid varieties __________
   (ii) Non hybrid varieties ______
   (iii) Bt-cotton __________
   (iv) *Desi* cotton ________
3. Total expenditure on irrigation
   (i) With tube well ___________ (Rs per acre)
   (ii) With canal water ___________(Rs per acre)

Insect Pest Control and Use of Insecticides

1. After how many days you usually visit your cotton field? ___________
2. (i) Do you apply insecticides after calculating ETL (Economic threshold level)? Yes/No
   (Please tick ☑ the correct one).
   (ii) If yes, for which insect pest you calculated the ETL ?

Sl.No.	*Name of Insect Pest*	*ETL (Economic Threshold Level)*
(*a*)	____________________	____________________
(*b*)	____________________	____________________
(*c*)	____________________	____________________

Contd...

Table 4.2–Contd...

(iii) If yes, in how much area did you spray insecticide after finding out the ETL? _____ (acre)

(iv) If no, what were the constraints in not using this practice (ETL)? Please Note down.

(*a*) ______________

(*b*) ______________

(*c*) ______________

3. Please note down the information regarding the use of insecticides in 2004 as per the following performa.

Sl.No.	*Date of Spray*	*Name of Insecticide Used*	*Quantity of Insecticide Used (per acre)*	*Quantity Water Used*	*Name of Insect for which Used*	*Area Sprayed (acre)*	*Variety of American/ Desi Cotton/ Bt Cotton*
1.							
2.							
3.							
4.							
5.							
6.							
7.							
8.							
9.							

2. Please note down the information regarding total quantity of different insecticides used.

Sl.No	*Name of Insecticide*	*Total Quantity Used (kg)*	*Cost (Rs.)*
1.			
2.			
3.			
4.			
5.			
6.			
7.			

3. Did you use hand sprayer or tractor mounted sprayer or both?

(Please tick ☑ the correct one)

(i) Hand sprayer

(ii) Tractor mounted sprayed

(iii) Hand plus tractorized sprayer (both)

Contd...

Table 4.2–Contd...

5. Cost of spraying one acre of land (Rs per acre)
   (ii) With hand sprayer Rs ___
   (ii) With tractorized sprayer Rs ____
6. In case of tractorized spryer, the quantity of diesel consumed for spraying one acre of land _______ (litre/acre)

Picking and Yield

1. Date of first picking ___________
2. Date of last picking ___________
3. Total pickings per acre ___________
4. Please note down the yield per acre and total production in case of different varieties.

Sl.No.	*Name of the Variety*	*Area (acre)*	*Yield per Acre*	*Total Production*
1.				
2.				
3.				
4.				
5.				

5. Total production of all the varieties ________________(in q)
6. Sale rate of cotton ________________ (in quintal)
7. Gross income from cotton crop ___________(in Rs.)
8. In your opinion, which was the best variety cultivated by you? ________
9. Cost of picking one quintal of cotton ___________ (in Rs.)

4.7.2.3 Content Area

The content area was IPM practices covered under IRM strategy in cotton crop. These were divided as under:

1. Different primary and secondary cultural practices, IRM strategy for insecticide use based on ETL and conservation of natural enemies.
2. Identification of insect pests, natural enemies from specimens.
3. Awareness and perception of the insecticides resistance in insect pests.

4.7.2.4 Preparation of Test Blue Print and Writing Test Items

Test items for knowledge test were written in consultation with the specialist from the IPM by giving emphasis to each content area of IRM strategy and type of items. The test contained open-ended questions and a few numbers of alternative response items. The statements of the items were written based on IPM practices for cotton crop enlisted under the Punjab Agricultural University (PAU) publication

"Insecticide Resistance Management in Cotton" (Dhawan *et al.*, 2004). In case of dosages of fertilizers, the package of practices of PAU was consulted. Content validity was ensured by consulting IPM specialists of PAU. Improvements were made based on their observations and suggestions. The knowledge test thus, constructed consisted of 30 items.

Table 4.3: Research Schedule 3–Adoptability Scale and Constraints

Department of Extension Education

Punjab Agricultural University

Ludhiana–141004

Title of the Research Problem: Evaluation of dissemination of insecticide resistance management technology in cotton crop in Punjab.

Sl.No. ________________

Date: ________________

1. Name of the farmer: ______________________________
2. Father's Name: ______________________________
3. Village: ______________________________
4. District: ______________________________

Part I: Adoptability and constraints in the adoption of selected IPM practices

1. ***Total area under cotton crop during 2004***________________acres

   (i) Area sown under cotton crop during April ________________acres

   (ii) Area sown under cotton crop after April ________________acres

   (iii) What were the considerations for sowing the cotton crop in April?________________

   (iv) What were the constraints for partially sowing the cotton crop in the month of April? ______________________

   (v) What were the constraints for not at all sowing the cotton crop in April?______________

2. ***Attributes of Timely Sowing***

   (*a*) *Relative Advantage*

   (i) Is there any relative advantage of April sowing? — <u>Relative Advantage</u> Complete/Partial/No

   (ii) What are/were the benefits of April sowing?
   - (*a*) Higher yield
   - (*b*) Less insect-pest attack
   - (*c*) Any other ____________

   (iii) Was the yield of April sown cotton crop higher than late sown? — Yes/No

   (*b*) *Compatibility*

   (i) Is April sowing compatible with your farming system/previous experience? — <u>Compatibility</u> Complete/Partial/No

   (ii) With respect to what it is compatible or non-compatible?
   - (*a*) Cultivation of preceding and succeeding wheat crop
   - (*b*) Size of land holding
   - (*c*) Irrigation facilities
   - (*d*) Any other ____________

Contd...

Table 4.3–Contd...

(*c*) *Observability*

(*i*) Are the benefits of April sowing observable (visible)?

<u>*Observability*</u>
Complete/Partial/No

(ii) What benefits of April sowing are/were observable ?

(*a*) More Yield
(*b*) Less insect-pest attack
(*c*) Any other ___________

(iii) Did April sown cotton crop give higher yield than may sown cotton crop?

Yes/No

3. ***Give the detail of area under different recommended resistant varieties of cotton crop during 2004.***

Hybrid	(*a*) LHH-144 ____________	acres
	(*b*) Ankur 651____________	acres
	(*c*) White gold ____________	acres
Non-hybrid	(*a*) F-1861 ____________	acres
	(*b*) LH-1556 ____________	acres
Any other	(*a*) ____________	acres
	(*b*) ____________	acres
	(*c*) ____________	acres

(i) What were the considerations for sowing resistant varieties mentioned above? ____________

(ii) What were the considerations for sowing other varieties? ____________

(iii) What were the constraints in not sowing the above mentioned resistant varieties? ____________

4. ***Attributes of Resistant Varieties***

(*a*) *Relative Advantage*

(i) Is there any relative advantage of cultivating the resistant varieties?

<u>*Relative Advantage*</u>
Complete/Partial/No

(ii) What are the relative advantages?

(*a*) Less insect-pest attack
(*b*) Less disease infestation (CLCV)
(*c*) Higher yielding
(*d*) Less pesticides use
(*e*) Higher profits
(*f*) Any other ___________

(b) *Observability*

(i) Are (were) the benefits of cultivating resistant varieties observable?

<u>*Observability*</u>
Complete/Partial/No

(ii) What benefits are observable?

(*a*) Less insect-pest attack
(*b*) Less disease infestation (CLCV)
(*c*) Higher yielding
(*d*) Less pesticides use
(*e*) Higher profits
(*f*) Any other ___________

Contd...

Table 4.3–Contd...

	Question	Response
	(*c*) *Compatibility*	
	(i) Is cultivation of recommended resistant varieties compatible with your farming system	Compatibility Complete/Partial/No
	(ii) With respect to what it is compatible or non-compatible?	(*a*) Compatible in your area (b) Compatible with irrigation facilities in your area (*c*) Compatible for picking (*d*) Any other__________
	(iii) Which variety (ies) of cotton you have planned (decided) to cultivate during 2005?	(1) __________ (2) __________ (3) __________
	(iv) What is your main criterion for selection of a variety?	__________ __________
5.	***Did you Treat Cotton-Seed Before Sowing?***	Yes Treated / No seed / Used Already
	(i) If yes, what were the reasons for seed treatment/sowing treated seed?	__________ __________
	(ii) If no, what were/are the constraints for not treating the cotton seed?	__________ __________
6.	***Attributes of Seed Treatment or Treated Seed***	
	(*a*) *Relative Advantages*	
	(i) Is there any benefit (advantage) of treating the cotton seed before sowing?	Relative Advantages Complete/Partial/No
	(ii) What are the benefits of seed treatment?	(*a*) Less jassid attack (*b*) Less disease attack (*c*) Any other __________
	(*b*) *Observability*	
	(i) Are the benefits of seed treatment observable	Observability Complete/Partial/No
	(ii) What benefits are observable?	(*a*) Apparent difference in the pest attack (*b*) Any other __________
	(*c*) *Complexity*	
	(i) Is seed treatment complex (difficult to adopt)?	Complexity Complete/Partial/No
	(ii) Complexity with respect to what?	(*a*) Difficult to treat the seed (*b*) Laborious (*c*) Skill intensive (*d*) Any other __________
	(*d*) *Compatibility*	
	(i) Is seed treatment/using treated seed compatible with other cotton growing practices?	Compatibility Complete/Partial/No
	(ii) With respect to what it is compatible or non-compatible?	__________

Contd...

Table 4.3–Contd...

7.	***Did you Spray the Pesticide Using ETL?***	Yes/No
	(i) If yes, what are (were) the apparent benefits of finding ETL?	______________ ______________
	(ii) If no, what are (were) the constraints of not finding ETL?	______________ ______________
	(iii) For which insect-pests you have the skills for finding ETL?	(*a*) Jassid (*b*) ABW (*c*) SBW (d) White fly
8.	***Attributes of Econmic Threshold Level (ETL)***	
	(*a*) *Relative Advantage*	
	(i) Is there any relative advantage of finding ETL before spraying of pesticide?	*Relative Advantage* Complete/Partial/No
	(ii) What are the advantages (benefits) of finding ETL?	(*a*) Judicious and need based use of pesticides (*b*) Less pesticide sprays (*c*) Saving on pesticide expenditure (*d*) Any other ______
	(b) *Observability*	
	(i) Are the benefits of finding ETL observable ?	*Observability* Complete/Partial/No
	(ii) What benefits of ETL are observable?	______________ ______________
	(*c*) *Complexity*	
	(i) Is finding ETL complex? (difficult to find out/use)	*Complexity* Complete/Partial/No
	(ii) What are the Complexities?	(*a*) Difficulty in observing the damage of a particular pest (*b*) Difficult to calculate (*c*) Labour intensive (*d*) Skill intensive (*e*) Any other ______
	(*d*) *Compatibility*	
	(i) Is finding ETL compatible with your pest management practices?	*Compatibility* Complete/Partial/No
	(ii) With respect to what it is compatible/ non-compatable?	(*a*) Your knowledge of ETL (*b*) Your skills (*c*) Your previous pest management practices

Contd...

Table 4.3–Contd...

9.	***Did you Apply Recommended Insecticides as per the Recommendations Given in IRM Literature***	Yes/No.
	(i) If yes,	(*a*) In full (*b*) partially
	(ii) What were the reasons for following IRM recommendations?	__________ __________
	(iii) If no, what were the constraints?	__________ __________
	(iv) Who took the decision regarding use of a particular insecticide	(*a*) yourself (*b*) your family member(s) (*c*) other farmers (*d*) Arhtia (*e*) IRM staff (*f*) Any other __________
	(v) What was the sources of information for selection/use of pesticides?	(*a*) IRM–Project staff (*b*) IRM–Literature (*c*) Other Literature of PAU (*d*) Other farmers (*e*) Pesticide companies (*f*) Arhtia (*g*) Radio (*h*) Television (*i*) Newspapers (*j*) Any other (specify) ________
10.	***Attributes of Insecticide Recommendation***	
	(*a*) *Relative Advantage*	
	(i) Is there any relative advantage (benefit) in following IRM recommendation for insecticide use ?	<u>*Relative Advantage*</u> Complete/Partial/No
	(ii) What are the benefits of adopting IRM recommendations?	(*a*) Less pesticide consumption (*b*) Reduced pesticide expenditure (*c*) Effective control of pests (*d*) Any other __________
	(*b*) *Observability*	
	(i) Are the benefits of following IRM schedule/ recommendations for use of pesticide observable?	<u>*Observability*</u> Complete/Partial/No
	(ii) What benefits are/were observable?	(*a*) Less expenditure of on pesticide (*b*) Effective control of pests (*c*) Higher yields (*d*) Any other __________

Contd...

Table 4.3–Contd...

(*c*) *Compatibility*

(i) Are the IRM recommendations for pesticide use compatible with your past experiences? — Compatibility: Complete/Partial/No

(ii) With respect to what it is compatible/non-compatible ? ____________

(*d*) *Complexity*

(i) Are the IRM-recommendations regarding use of pesticides difficult to adopt? — Complexity: Complete/Partial/No

(ii) What are the complexities ? ____________

Part II: Farmers Reactions About IRM-Project

1. ***What are your observations about the IRM-project which was implemented in your village?*** ____________

2. ***Was the IRM-Technology of any benefit to you?*** Yes/No
   (i) If yes, what were the benefits of this programme? ____________
   (ii) If no, what were the reasons for your dissatisfaction? ____________

3. ***Were you regular in attending the group meetings?*** Yes/No
   (*i*) How many group meetings did you attend? (approximately) ____________
   (ii) If no, what were the reasons? ____________

4. ***Did you get the IRM-booklet?*** Yes/No
   (i) If yes, is the information contained in the booklet
   (*a*) Useful for cotton cultivation Yes/No
   (*b*) Language of the literature understandable Yes/No
   (*c*) Comprehensive Yes/No

5. ***In your opinion what should be done to improve the benefits of IRM programme to the farmers with respect to:***
   (i) Training ____________
   (ii) Literature ____________
   (iii) Field observation ____________
   (iv) Experimentation ____________
   (v) Timing of group meetings ____________
   (v) Any other suggestion ____________

Contd...

Table 4.3–Contd...

6.	***Did you purchase pesticide on***	*Cash/credit ?*
	(i) In case of credit did the pesticide dealer/*Arhtia*-influence you for purchasing a particular insecticide?	Yes/No
	(ii) In case of cash purchase did the pesticide dealer/ *Arhtia* influence you for purchasing a particular insecticide?	Yes/No
	(iii) Were you given the bill on the purchase of insecticide?	Yes/No
7.	***Are you satisfied with the IRM-Trainers with respect to their?***	(*a*) Communication Yes/No (*b*) Technical capabilities Yes/No (*c*) Behaviour Yes/No (*d*) Punctuality Yes/No (*e*) Conducting of group meetings Yes/No
8.	***Were you facilitated/imparted training in the identification of insect-pests and natural enemies?***	Yes/No
9.	***Were you practically imparted skills to calculate ETL?***	(*a*) Jassid (*b*) White fly (*c*) Boll worms
10.	**What you learned from the street play staged in your area?**	________________ ________________
11.	***Have you heard about Bt-cotton***	Yes/No
	(i) If yes, what was the source of information?	(*a*) Television (*b*) Radio (*c*) Other farmers (*d*) Companies (*e*) *Arhtia* (*f*) Newspapers (*g*) PAU staff (*h*) Agri. Department (ADO) (*i*) Any other (specify____________
	(ii) In which year did you cultivate the *Bt*-cotton	________________
12.	***Did you cultivate Bt-cotton during 2004?***	Yes/No
	(i) Was the source of seed authentic?	Yes/No
	(ii) What were the reasons for cultivating *Bt*-cotton?	(*a*) Less pest attack (*b*) Higher yields (*c*) Saving on pesticide expenditure (*d*) Easy to adopt (*e*) Compatible with your farming (*f*) Could be experimented with (*g*) Any other ____ _______
(iii)	How much area you plan to bring under *Bt*-cotton during 2005?	________________ acres

Contd...

Table 4.3–Contd...

13. ***Give the detail about the number of pesticide application during Kharif 2004 in the following crops/varieties***

Sl.No.	Crop/Variety	Total no. of Pesticides Sprays	Against which Pest	No. of Sprays as Mixture	Expenditure on Pesticide (Rs.)	Yield (acre)
1.	*Bt*-cotton					
2.	Other cotton varieties					
3.	Rice varieties					
	Area: ___ acres					
	Varieties:					
	(i)					
	(ii)					
	(iii)					
4.	Other crops					
	(i)					
	(ii)					
	(iii)					

4.7.3 Construction of Questionnaire

The second research schedule was in the form of semi-structured questionnaire. The questionnaire was to study IPM practices adopted for cotton cultivation. This questionnaire was in Punjabi and related to:

1. Extent and level of adoption of different cotton growing practices by the respondent farmers in IRM and non-IRM villages and use of pesticides per acre.
2. Input-output cost of cotton cultivation for calculating cost of cultivation, gross returns, and proportion of pesticides cost to total cost of cultivation per hectare.

4.7.4 Construction of Research Schedule for Measuring Adoptability

The research schedule number three consisted of two parts.

Part I: To find out the reasons for and constraints in the adoption of selected IPM practices *viz.* adoption of timely sowing, recommended resistant varieties, seed treatment, ETL, following IRM strategy for proper use of insecticides. This part also included a three-point continuum scale for measuring the adoptability of the above-referred selected IPM practices.

Part II: The part second of this schedule was for evaluation of IRM training programme with respect to trainers, subject matter, training methods, training aids and to record opinion and suggestions of IRM farmers about the programme and ways for making it more effective. The open ended and semi-structured type questions were used to provide only leads to get comprehensive insight about the IRM programme from trainees' perspective. This part did not consist of forced-choice items.

4.7.4.1 Construction of the Scale

For construction of the scale for measuring the adoptability of selected attributes, an objective scale as discussed by Kerlinger (1973) was followed. In this type of the scale, subject responds to each item freely with a range of two or more possible responses from which a person can choose one.

4.7.4.2 Content Area

The content area for the scale was selected IPM practices *viz.* timely sowing of cotton crop, cultivation of recommended resistant varieties/hybrid varieties of cotton, seed treatment/treated seed, use of economic threshold level for cotton insect pests and use of insecticides as per IRM strategy. These practices were selected after discussion with experts from the field of entomology and extension education.

4.7.4.3 Selection of Attributes

The five attributes of innovations as listed by Rogers (1983) which are universally relevant were selected to measure adoptability of the selected IPM practices disseminated under IRM programme. These included relative advantage, compatibility, complexity, trialability and observability. The selection of attributes for a particular practice was done after discussions with the experts in the field of extension education. The attributes selected for a particular practice were:

Sl.No.	*Practice*	*Attributes*
1.	Timely sowing of cotton crop	Relative advantage, observability, compatibility, trialability
2.	Recommended resistant varieties of cotton	Relative advantage, observability, compatibility and trialability
3.	Seed treatment	Relative advantage, observability, compatibility, trialability and complexity
4.	Use of ETL	Relative advantage, observability, compatibility trialability and complexity
5.	Use of insecticides according to IRM strategy	Relative advantage, observability, compatibility trialability and complexity

4.7.4.4 Response Categories

Three-point response continuum as already discussed under sub section 4.5.6 was adopted. In addition to the scale, open-ended semi structured questions were also formulated to study the IRM farmers' response towards the selected attributes, so that any ambiguity is removed, and to validate the response given on three point continuum scale.

4.8 Pre-testing of Research Schedules

The research schedule number one- knowledge test and research schedule number three–adoptability scale and its other parts were pre-tested in non-sampled IRM village of district Mansa, covered under IRM programme in 2003 to find out the ambiguity, weakness in the items of format of, knowledge test items, statements of the scale and reasons and constraints part of the schedule. The research schedule for studying adoption and input-output cost, was also pre-tested in non-sampled village and at the time of distribution of questionnaire to the respondent farmers, the details were explained and were asked to point out, if there was any ambiguity. No changes were made in this schedule after pre-testing.

On the basis of experience gained and information obtained through pre-testing, necessary modifications were made in knowledge test and scale. Few items were deleted and few items were added to make it easy for recording the data, and to remove ambiguity.

In the blue print of knowledge test, there were 30 test items. After pre-testing, five items were deleted, and four items were reformulated after consulting the experts from the field of entomology to ensure content validity. Final knowledge test consisted of 29 test items. No discrimination or difficulty indices were computed for knowledge test as none of the test items could have been deleted. All the test items were important under the IRM strategy as envisaged in the IRM publication of PAU.

The research schedule to measure adoptability was modified after pre testing. In the initial scale, there were only two response categories; yes and no. After pre-testing the response categories were reformulated and third response category was also added. The response categories thus were; completely, partially and no. The trialability attribute was also deleted as the respondents in pre-testing reported all the practices as trialable.

4.9 Reliability of Knowledge Test and Scale

Test-retest method was used to determine the reliability coefficient. Problem was encountered while administering the same knowledge test again to previously administered respondents, as the respondents showed their reluctance to provide the same information again, because they were not taken into confidence that the same test was to be administered again. This experience gained helped at the time of determining the reliability of the scale by test-retest method. The correlation coefficient of test-retest was worked out.

$$r_{xy} = \frac{N\Sigma XY - (\Sigma X)(\Sigma Y)}{\sqrt{[N\Sigma X^2 - (\Sigma X)^2][N\Sigma Y^2 - (\Sigma Y)^2]}}$$

where,

r: Coefficient of correlation,

X and Y: Test and retest knowledge scores, and

N: Number of respondents

The reliability coefficient of knowledge test worked out 0.86 and reliability coefficient of adoptability scale worked out 0.79. The test therefore was considered reliable. In case of adoptability scale reliability coefficient being 0.79 may be because of the fact that only two point continuum was used in pre-testing.

4.10 Weightage Scheme for Knowledge Test Items

Differential scoring pattern was employed for 29 test items of knowledge test depending on the importance of an item (Table 4.1). Weight-age scheme was decided after consulting the members of the advisory committee from the field of extension education and entomology.

4.11 Data Collection

Data were collected in three stages from April 2004 to February 2005.The data related to profile of the respondents, base line data and pre-training knowledge test was collected by personal interview method at the start of cotton growing season in 2004. For knowledge test, real objects *viz.* insect pests, natural enemies and photographs of damage symptoms of insect pests were used to test the knowledge of respondents in IRM villages. The knowledge test was administered to 150 farmers (only in IRM villages). Post-training the knowledge test was again administered in IRM villages to find out the gain in knowledge.

The data related to adoption of different IPM practices disseminated under IRM programme were collected through a questionnaire, which was given to the respondent farmers at the start of cotton growing season in 2004 in both the IRM villages and non-IRM villages. Farmers were requested to record the weekly cotton cultivation operations carried out and inputs used in the questionnaires which was given in folders to each respondent farmer. These questionnaires were collected after the cotton growing season was over, in the months of December 2004 and January 2005. Out of 210 respondents, in IRM and non-IRM villages, 207 respondents returned the duly completed questionnaires.

To study the adoptability, constrains in the adoption of selected IPM practices and evaluation of IRM programme by farmers, data were collected by personal interview method from November 2004 to February 2005 in IRM villages only. Out of 150 farmers selected in IRM villages only 146 respondents could be interviewed as four respondents showed their reluctance.

The training programme under the IRM project starting in April 2004 was observed in selected villages from time to time up to September 2004. The observations were recorded in a diary.

Time series secondary data about cotton cultivation, use of pesticides, yield per unit area were collected from different sources. The data regarding the spread of the IRM project in India and Punjab were collected from, the State Coordinator, IRM Project, Punjab.

4.12 Objective Wise Research Hypotheses

Hypotheses were framed on different aspects of the study. Statistical formulations have been given for the hypotheses which have been tested by appropriate statistics.

4.12.1 Objective Number One

(*i*) Null Hypothesis (H_0)

There is no significant difference in pre and post-training mean knowledge score of IRM farmers

(*ii*) Alternative Hypothesis (H_1)

The post-training mean knowledge score is significantly higher than pre-training mean knowledge score of IRM farmers

$$H_0 : \overline{X}_1 = \overline{X}_2$$

$$H_1 : \overline{X}_2 > \overline{X}_1$$

where,

$\overline{X}_1$: Mean pre-training knowledge score of IRM farmers

$\overline{X}_2$: Mean post-training knowledge score of IRM farmers

4.12.2 Objective Number Two

(1) (*i*) Null Hypothesis (H_0)

There is no significant difference in the average insecticide applications by IRM farmers before and during the IRM programme

(*ii*) Alternate Hypothesis (H_1)

There is significant reduction in the average insecticide applications by IRM farmers during the IRM programme than before

$$H_0 : \overline{X}_1 = \overline{X}_2$$

$$H_1 : \overline{X}_2 < \overline{X}_1$$

where,

$\overline{X}_1$: Is the average insecticides applications before the IRM programme

$\overline{X}_2$: Is the average insecticides applications during the IRM programme.

(2) (*i*) Null Hypothesis (H_0)

There is no significant difference in the average insecticide applications between IRM and non-IRM farmers during 2004.

(*ii*) Alternative Hypothesis (H_1)

The average insecticide applications by IRM farmers are significantly lower than non-IRM farmers during 2004

$$H_0 : \overline{X}_1 = \overline{X}_2$$

$$H_1 : \overline{X}_1 < \overline{X}_2$$

where,

$\overline{X}_1$: Is the average insecticide applications by IRM farmers during 2004

$\overline{X}_2$: Is the average insecticide applications by non-IRM farmers during 2004

(3 (*i*) Null Hypothesis (H_0)

There is no significant difference in the average insecticide expenditure before and during the IRM programme.

(*ii*) Alternative Hypothesis (H_1)

There is a significant reduction in the average insecticide expenditure during the IRM programme than before

$$H_0 : \overline{X}_1 = \overline{X}_2$$

$$H_1 : \overline{X}_2 < \overline{X}_1$$

where,

$\overline{X}_1$: Is the average insecticide expenditure before IRM programme (2003)

$\overline{X}_2$: Is the average insecticide expenditure during IRM programme (2004).

(4) (*i*) Null Hypothesis (H_0)

There is no significant difference in the average insecticide expenditure between IRM farmers and non-IRM farmers during 2004

(*ii*) Alternative Hypothesis (H_1)

The average insecticide expenditure of IRM farmers is significantly lower than non-IRM farmers during 2004

$$H_0 : \overline{X}_1 = \overline{X}_2$$

$$H_1 : \overline{X}_1 < \overline{X}_2$$

where,

$\overline{X}_1$: Is the average insecticide expenditure of IRM farmers

$\overline{X}_2$: Is the average insecticide expenditure of non-IRM farmers.

(5) (*i*) Null Hypothesis (H_0)

There is no significant difference in the seed cotton yield per unit area between IRM and non-IRM villages during 2004

(*ii*) Alternative Hypothesis (H_1)

The seed cotton yield per unit area of IRM villages is higher than non-IRM villages during 2004

$$H_0 : \overline{X}_1 = \overline{X}_2$$

$$H_1 : \overline{X}_1 > \overline{X}_2$$

where,

$\overline{X}_1$: Is the seed cotton yield per unit area in IRM villages

$\overline{X}_2$: Is the seed cotton yield per unit area in non-IRM villages

(6) (*i*) Null Hypothesis (H_0)

There is no significant difference in average tank mixing of insecticide applications between IRM and non-IRM farmers during 2004.

(*ii*) Alternative Hypothesis (H_1)

The average tank mixing of insecticide applications of IRM farmers are significantly lower than non-IRM farmers during 2004.

$$H_0 : \overline{X}_1 = \overline{X}_2$$

$$H_1 : \overline{X}_1 < \overline{X}_2$$

where,

$\overline{X}_1$: Is is the average tank mixture of insecticide applications by IRM farmers

$\overline{X}_2$: Is is the average tank mixture of insecticide applications by non-IRM farmers

The hypothesis which was examined from the pattern, which emerged from the tabulated data but could not be tested by appropriate statistical methods, is:

(7) (*i*) Null Hypothesis (H_0)

There is no difference in insecticide consumption per ha (technical grade material) between IRM and non-IRM villages.

(*ii*) Alternative Hypothesis (H_1)

The insecticide consumption per ha (technical grade material) is less in IRM villages than non-IRM villages.

4.13 Statistical Analysis

The data were analyzed with the help of frequencies, percentages, mean, standard deviation, paired 't' test, two sample 't' test of unequal variance, indices and Pearson's correlation coefficient. Computer based programme SPSS (Statistical package for social sciences) was used for applying different statistical tests for finding out association and significance of means.

4.13.1 Standard Deviation (SD)

Standard deviation was computed to determine deviation from mean in case of average number of pesticide applications, hoeings, pesticide expenditure, irrigation, average dosages of fertilizers per hectare, yield per unit area etc.

$$S.D. = \sqrt{\frac{\sum (X - \overline{X})^2}{n-1}}$$

4.13.2 Paired 't' Test

Paired 't' test was applied to compare the mean difference between two paired observations to test whether the paired sample means differ from one another significantly in their mean. It was applied for comparing the 'pre and post' training mean knowledge scores; average pesticide applications, pesticides expenditure during 2003 and 2004. It was applied for testing hypothesis number 1 of 4.12.1 and for hypothesis number 1 and 3 of 4.12.2.

$$t = \frac{\overline{d} - 0}{S} \times \sqrt{n}$$

$$S = \sqrt{\sum \frac{(di - \overline{d})^2}{n-1}}$$

$$\overline{d} = \frac{\sum di}{n}$$

where,

$\overline{d}$: The mean of the difference

S: The standard deviation of the differences

d.f. = (n-1)

4.13.3 Two Sample 't' Test with Unequal Variance

Two-sample 't' test was applied to compare two means from two independent samples to test whether the two samples differ from one another significantly in their mean. It was applied for testing the hypothesis number 2, 4, 5 and 6 of 4.12.2.

$$t = \frac{\overline{X}_1 - \overline{X}_2}{\sqrt{\frac{S_1^2}{n_1} + \frac{S_2^2}{n_2}}}$$

$$d.f = \frac{\left(\frac{S_1^2}{n_1} + \frac{S_2^2}{n_2}\right)^2}{\sqrt{\frac{\left(\frac{S_1^2}{n_1}\right)^2}{(n_1 - 1)} + \frac{\left(\frac{S_2^2}{n_2}\right)^2}{(n_2 - 1)}}}$$

where,

X_1 : Mean of the first sample

$\overline{X}_2$: Mean of the second sample

d.f: Degree of freedom

S_1: Standard deviation of I sample

S_2: Standard deviation of II sample

4.13.4 Karl Pearson's Coefficient of Correlation

Pearson's correlation coefficient was applied to find out the degree of association between education and knowledge, land holding and number of insecticide applications and education.

$$r_{xy} = \frac{N\Sigma XY - (\Sigma X)(\Sigma Y)}{\sqrt{[N\Sigma X^2 - (\Sigma X)^2][N\Sigma Y^2 - (\Sigma Y)^2]}}$$

where,

r: Coefficient of correlation

X and Y: Variables

N: Number of observations.

Chapter 5

EVALUATION OF THE INSECTICIDE RESISTANCE MANAGEMENT PROGRAMME

Keeping in view the objectives of the study, the results and discussion have been elaborated under the following heads:

5.1 Selection procedure of villages and farmers under IRM programme
5.2 Profile of the respondents
5.3 Extension methodology followed for dissemination of technology
5.4 Knowledge gain of the farmers about IPM
5.5 Adoption of IPM practices
5.6 Economic benefits at the farmers' level
5.7 Adoptability and constraints
5.8 *Bt*-cotton: adoption before recommendation
5.9 Evaluation of the IRM programme by the trainee farmers
5.10 Inferences

5.1 Selection Procedure of Villages and Farmers Under IRM Programme

Before the start of cotton growing season in 2004, the district level trainers of the IRM project made the selection of villages covered under the programme. The selection of villages was based on the views of the local scouts and agricultural development officers of the state department of agriculture. The meetings of farmers were organised in the short listed villages to enlist the views of the *'Sarpanches'* and farmers for conducting the IRM programme in their villages. In case of positive response of the *Sarpanch* and farmers, the village was finalized for inclusion in IRM programme. The villages nearby the selected IRM villages were selected as control villages under the IRM programme.

In both, the IRM and non-IRM villages a list of cotton growers was prepared by the IRM staff with the help of local *'Sarpanch'*. The farmers in IRM villages included in the list formed the cotton growers group for weekly meetings and other training activities. Out of these farmers, a core group of ten farmers was selected in each IRM village by the district level trainers. The farmers selected were those who were active farmers and who volunteered to actively and regularly participate in different activities of the IRM programme. In certain villages more farmers showed keen interest to be the part of the core group, but only 10 farmers were randomly taken by the trainers for the purpose of monitoring insect pest resistance in the field and other related activities. The farmers selected from IRM villages from now on will be referred to as IRM farmers and from non-IRM villages will be referred to as non-IRM farmers. The sampling procedure for selection of villages and respondents for this study has been discussed in Chapter 4.

5.2 Profile of the Respondents

The comparison of IRM farmers with non-IRM farmers in the 15 IRM villages and six non-IRM villages studied is based on the data collected in the baseline survey prior to IRM programme in 2004. The data of comparison is given in Table 5.1.

5.2.1 Educational Level

The majority of the respondents, in IRM and non-IRM villages were literate. The data reveals that 10 per cent of the respondents in IRM villages, and 15 per cent of the respondents in non-IRM villages were illiterate. Except in case of district Bathinda, where there were more illiterates in non-IRM villages (20 per cent) compared to IRM villages (6 per cent), there was negligible difference in other two districts of Ferozepur and Mansa. Majority of the respondents (70 per cent) were having education from middle to 10+2, in both the IRM and non-IRM villages. This shows that the sample of IRM farmers and non-IRM farmers had similar educational level.

5.2.2 Operational Land Holding

The average farm size of the farmers varied from district to district (Table 5.1). In Bathinda, the average holding size was 9.22 ha in case of IRM farmers and 5.56 ha in case of non-IRM farmers. In Mansa the average holding size was 7.54 ha and 5.51 ha, in case of IRM and non-IRM villages, respectively. But in district Ferozepur, the average holding size of both IRM and non-IRM farmers was 13.50 ha and 15.62 ha, respectively. In district Ferozepur the holding size is more compared to other two districts. Overall, the average holding size was 10.09 ha and 8.90 ha in case of IRM and non-IRM farmers, respectively.

A small percentage of four farmers, in IRM villages belonged to small farmers having holding size 1-2 ha, where as 13 per cent, of non-IRM farmers belonged to this

Table 5.1: Profile of the Respondent Farmers

Particular	District Bathinda		District Ferozepur		District Mansa		Overall for Three Districts	
	IRM Villages	Non-IRM Villages	IRM Villages	Non-IRM Villages	IRM Villages	Non-IRM Villages	IRM Villages	Non-IRM Villages
Education (% farmers)								
i. Illiterate	6.00	20.00	8.00	10.00	16.00	15.00	10.00	15.00
ii. Up to primary	10.00	0.00	18.00	15.00	12.00	10.00	13.00	8.00
iii. Middle	18.00	25.00	26.00	15.00	26.00	25.00	23.00	22.00
iv. Matric	36.00	40.00	36.00	50.00	32.00	45.00	35.00	45.00
v. 10+2	14.00	0.00	8.00	5.00	14.00	5.00	12.00	3.00
vi. Graduate and above	16.00	15.00	4.00	5.00	0.00	0.00	7.00	7.00
Telephone connection (% farmers)	84.00	30.00	66.00	60.00	64.00	35.00	71.00	42.00
Total operational land holding (ha) (i + ii - iii)	461.20	111.20	675.00	312.40	377.00	110.20	1513.20	533.80
i. Owned	415.80	93.60	611.80	287.60	352.40	96.20	1380.00	477.40
ii. Leased-in	53.00	17.60	74.80	24.80	29.60	14.00	157.40	56.40
iii. Leased-out	7.60	0.00	11.60	0.00	5.00	0.00	24.20	0.00
Average operational land holding (ha)	9.22	5.56	13.50	15.62	7.54	5.51	10.09	8.90
Operational land holding (% farmers)								
i. 1–2ha (small)	6.00	25.00	4.00	0.00	2.00	15.00	4.00	13.00
ii. 2–4ha (Semi-medium)	20.00	20.00	12.00	10.00	24.00	35.00	19.00	22.00
iii. 4–10ha (Medium)	46.00	20.00	40.00	30.00	56.00	40.00	47.00	30.00
iv. >10ha (Large)	28.00	35.00	44.00	60.00	18.00	10.00	30.00	35.00
Area under cotton crop during 2004 (ha)	313.00	75.80	428.20	208.40	217.65	66.40	958.85	350.60
Percentage area under cotton crop during 2004	67.87	68.17	63.44	66.71	57.73	60.25	63.37	65.68

Percentage of farmers rounded up to the nearest whole number.

group. The selection of more percentage of small farmers from non-IRM villages for the present study must be due to random sampling technique adopted for selection of non-IRM respondents which was not the case in IRM villages. In case of IRM farmers, the sample selection was purposive, as there was no possibility of random selection. But the majority of the farmers 77 per cent and 65 per cent, in IRM and non-IRM villages, respectively, were either medium or large farmers. The percentage of area under cotton crop during 2004 was 63.67 and 65.68 per cent in IRM and non-IRM villages, respectively. The villages selected in each district were predominantly cotton growing and it is the main *Kharif* season crop of this belt of Punjab.

The comparison of IRM and non-IRM farmers with respect to productivity of cotton crop, average pesticide applications and expenditure per season collected in the base line data has been discussed under the sub section 5.6.4 of this chapter.

5.3 Extension Methodology Followed for Dissemination of Technology

The efforts to disseminate IPM through transfer of technology (ToT) model failed (Kenmore *et al.*, 1995, van de Fliert 1993). ToT model was successful in disseminating the green revolution technologies like high yielding varieties, fertilizers etc. This model is successful as long as the technology is simple and not interactive. The field evaluation of IPM programmes in areas where ToT model was applied showed that pest population in some of the IPM areas had increased in 1984 (van de Fliert 1993). IPM programmes focusing on transfer of specific technologies or bits of information would not succeed. In Punjab, even after the introduction of many IPM programmes the average number of pesticide applications as reported by the farmers covered under IPM programmes earlier had increased in some of the study villages.

In Punjab, after the introduction of various IPM programmes in cotton, the insecticide cost to total cost of cultivation increased from 11.9 per cent in 1984-85 to 15.5 per cent in 1989-90 to 13.0 per cent 1994-95 (Dhaliwal and Arora 2001) to 21.2 per cent in 1998-99 (Sen and Bhattia 2004) to around 50 per cent (Shetty 2004). However, during 1998 the cost of insecticides used in the state was higher than that of the productivity and farmers got negative returns from cotton cultivation. The Farmer Field School model seen as the classical model in the developing nations has not shown desired results in Punjab, if the above referred trends are any indicator. This was due to the development of resistance to insecticides in cotton pests particularly American bollworm, and lack of following the extension methodology associated with FFS, as in India it is same methodology of instructing farmers, which was used before FFS in 1991 (Ooi *et al.*, 1992) to 1995-96 (FAO 1996).

Therefore, the extension methodology followed in IRM programme has been evaluated in that context. According to Russell (2004) "IRM provided a different IPM model from that of FFS. Much of the same information–pest and beneficial identification, scouting practices, appropriate use of chemistries was provided in a more didactic form, albeit through active participation throughout the season. The messages disseminated from scout to farmer relatively easily, but this does not provide the farmers with the experience in participatory research that would help them to cope with a significant changed pest context." The IRM programme is directly being implemented by the PAU scientists from the field of entomology. Extension methodology adopted for dissemination of IRM strategy is given in Table 5.2.

Table 5.2: Extension Methodology Followed for Dissemination of IPM Practices Under IRM Programme

1. **Training approach**	(*i*)	Farmers groups in IRM villages
2. **Training methods**	(*i*)	Meetings starting from sowing of cotton crop.
	(*ii*)	Group meetings every week: July to end of September.
	(*iii*)	Training by master trainer for each of the four windows of the IRM strategy for judicious use of insecticides.
	(*iv*)	Visit to IRM laboratories set up at PAU research stations.
	(*v*)	Visit to the selected good IRM farmer's field.
	(*vi*)	Lectures and discussions
	(*vii*)	Trainers as experts, not the facilitators of learning process
	(*viii*)	Training of scouts before the start of project and during 2003
	(*ix*)	Scouts deployed in every village for providing feed back, estimation of pest build up and organizing meetings every week.
	(*x*)	Visit to farmers' fields for identification of insect pests, natural enemies and other problems.
	(*xi*)	Information centres established at IRM villages
3. **Training tools**	(*i*)	Field training
	(*ii*)	Publication and distribution of IRM printed material to farmers
	(*iii*)	Posters, banners, pamphlets, display panels, specimens
	(*iv*)	Street plays organized for awareness and for creating interest.
	(*v*)	Use of public address system for dissemination of information

5.3.1 Farmers Meetings

The IRM training programme lasted for whole cotton growing season. From sowing until 60 days after sowing, there were occasional farmers' group meetings in all the IRM villages. July onwards, the training programme gained in intensity with weekly group meetings in the villages up to end of September. Trainers acted as information providers to the farmers. At the time of sowing cotton crop, farmers were advised to cultivate those varieties which were jassid resistant, had resistance against cotton leaf curl virus (CLCuV) and were early maturing (maturing in 160-170 days after sowing), otherwise they were advised to cultivate PAU recommended varieties. The main emphasis of group meetings was on delaying early application of insecticides, as far as possible up to 90 days after sowing.

In the weekly group meetings the problems faced by the farmers since the last meeting were discussed and the expert information was provided for solving the problems. The training and other information on weekly basis was provided by senior research fellows (specialists in the field of plant protection) incharge of a district. Expert information was provided by the scientist (master trainer) from the field of entomology before the beginning of each window of IRM strategy.

5.3.2 Deployment of Field Scouts

Field scouts were deployed, one for every village and the scouts were mostly local farmers. The scouts acted as a link between the project/university staff and the farmers. The scouts used to monitor pest problem for the whole week and regularly communicate the information to district level scientists looking after the programme.

5.3.3 Training Activities

The training activities besides group meetings included the printed information provided on the status of insect pests, natural enemies, ETL and pest biology. Wide array of IPM teaching activities such as technical bulletins, crop/pest guidelines in the form of manual, field days, tours of IRM farmers to IRM laboratories set up at district headquarters were adopted. Information Centres were set up at the IRM villages and normally weekly group meetings were conducted there. Award winning street play '*Sambal Samaj Kar Hee Keet Nashak ka Upyog-Ek Mukti Marg*' (Judicious use of pesticides-a way out of pesticide treadmill) was organised at IRM villages and district headquarters by the PAU students to create awareness about the adverse effects of over reliance and indiscriminate use of pesticides. Photographs of different extension and other activities carried out during the programme are shown in Figures 5.1 to 5.10.

Figure 5.1: Pre-seasonal Meeting with Farmers

Figure 5.2: Group Meeting of IRM Farmers

Figure 5.3: Identification of Pests and Natural Enemies in Farmer's Field

Figure 5.4: Training Session of IRM Farmers with Master Trainer

(a) (b) (c)

Figure 5.5: Printed Literature Provided to Farmers in IRM Village
(a) Leaflet; (b) IRM manual; (c) Photographs of pests and natural enemies of cotton

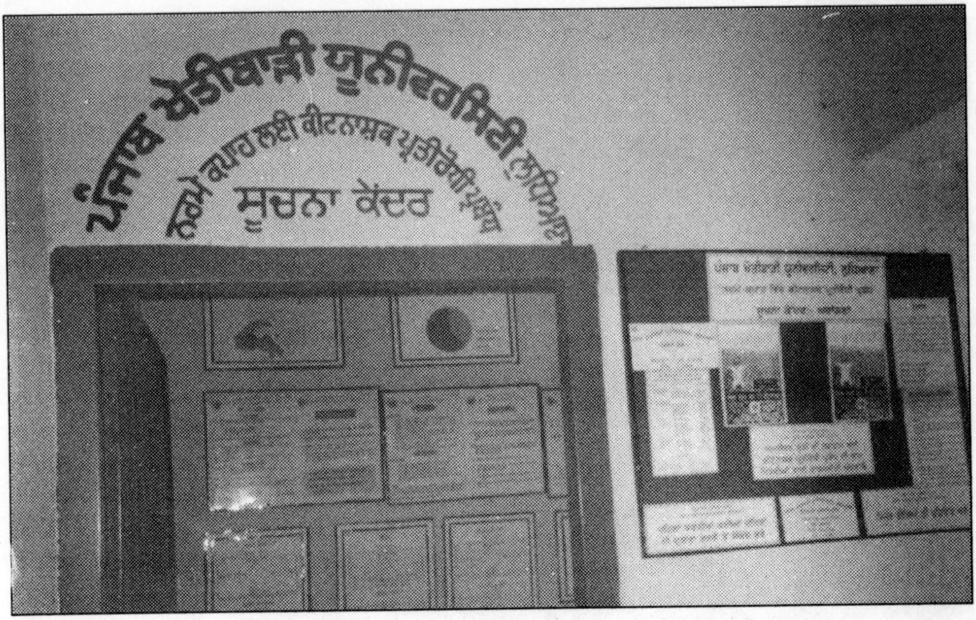

Figure 5.6: Information Centre at IRM Village

Figure 5.7: PAU Students Performing Street Play at IRM Village

Figure 5.8: Visit of IRM Farmers to PAU Research Station at Abohar

Figure 5.9: Visit of IRM Farmers to IRM Laboratory at KVK, Bathinda

Figure 5.10: Scientists from CICR, Nagpur Interacting with IRM Farmers

5.3.4 A Critical Analysis of Training Strategy

With the experience of having been associated with FAO-ICP for IPM conducted by Directorate of Plant Protection Quarantine and Storage, Government of India in association with state agricultural departments, the FFS model mainly laid emphasis on identification of pests and their natural enemies and avoidance of use of chemicals. There was less emphasis on the pest management strategies in FFS. In the beginning, during 1994 the programme was run by master trainers and slowly passed into the hands of trainers who were not comfortable with IPM strategy. As the programmes scaled in quantity, the quality of training deteriorated. The trainers lacked competence in the content area as well as process area (Peshin and Kalra 1998). No worthwhile coordination was observed between the scientists from the universities and state departments of agriculture (Peshin and Kalra 2000a). The Directorate of Plant Protection, Quarantine and Storage, Government of India is conducting IPM programmes independently with a limited reach.

In the IRM programme, the emphasis is mainly on manipulation of cultural practices and judicious use of insecticides in the absence of other strategies available, though presented within full IPM context. The training responsibility has been directly given to state agricultural universities and in the hands of specialists from the field of entomology. This way the quality information being disseminated is credible with continuous monitoring of pest build up and insecticide resistance. Though the extension methodology has been termed participatory in the IRM guidelines, but the participation was passive in terms of farmers being told what was going to happen or has already happened and limited to participation by consultation. Participation by consultation implies that people participate by being consulted by trainers, but trainers define both problem and solution and modify these in the light of farmers' responses (Pretty 1994). The IRM programme has the following shortcomings:

1. No on farm result demonstrations laid the results of which could be viewed by participating farmers 'as seeing is believing'.
2. More emphasis was given to lecturing, instead of on-field experiential learning. The field should be the laboratory for farmers for acquiring knowledge and skills.
3. The programme takes off slowly at the sowing time, it should continue from seed to seed with the same intensity as was observed during the four windows of IRM strategy.

The strategy adopted to aggressively compete with pesticide agents and companies through wall paintings, posters in villages, mass media and providing the IRM farmers' identity cards were the novel methods employed in IRM programme. This had a tremendous impact during 2004. The evaluation of the programme from farmers' perspective is discussed at the end of this chapter under sub section 5.9.2.

5.4 Knowledge Gain of the Farmers about IPM

The knowledge of IRM farmers about different IPM practices selected under the programme was measured on 29-item knowledge test constructed for the purpose of the study. Pre and post-training, percentage of farmers having knowledge about selected aspects of IPM was compared to find out how much percentage of farmers benefited from the programme with respect to knowledge gain. In non-IRM villages knowledge test was not administered.

The practical knowledge of the farmers about the characteristics of a healthy crop, identification of major pests, infestation symptoms, natural enemies, interaction of pests and natural enemies, life cycle of pests and familiarity with practices that prevent damage and control/manage pest population are the knowledge requirement of farmers to be characterised as having good knowledge of IPM (van de Fliert 1998) and which can lead to further action on the part of the farmer. All these aspects were covered in the knowledge test constructed for the study.

5.4.1 Knowledge About Recommended Resistant Varieties

The Punjab Agricultural University (PAU) has recommended varieties, which are resistant to CLCuV. The sowing of CLCuV susceptible varieties is strongly discouraged in Punjab in CLCuV hotspots. Under the IRM programme the resistant/ tolerant varieties recommended were LHH-144, Ankur 651, White gold (hybrid cotton); F 1861 and LH 1556 (non hybrid varieties). Out of these, Ankur 651 and White gold are not developed by the PAU but are recommended by the university.

Before the programme, 35 and 11 per cent of the IRM farmers had knowledge about Ankur 651 and White gold hybrid varieties, and only 11 per cent about LHH 144. Some farmers were not aware that Ankur 651 is a PAU recommended variety. The knowledge about Ankur 651 was more among farmers in district Bathinda and Mansa in comparison to district Ferozepur. In the first two districts the adoption of hybrid varieties is many folds higher than in Ferozepur district, so percentage of farmers having knowledge about recommended hybrid varieties was higher. After the IRM programme, the percentage of farmers having knowledge about the PAU recommended resistant varieties almost doubled. Before and after comparison with respect to percentage of farmers having knowledge about recommended resistant varieties is given in Table 5.3. The percentage of farmers having knowledge about the recommended hybrid varieties increased by 10-27 per cent and the percent change was more in districts Bathinda (8-32 per cent) and Mansa (8-30 per cent) in comparison with district Ferozepur (6-18 per cent) (Table5.3). But in case of non-hybrid resistant varieties the increase in percentage of farmers having knowledge was more in Ferozepur (30-32 per cent) in comparison to Mansa (4-10 per cent) and Bathinda (6 per cent). This shows that farmers are keen on acquiring knowledge about those practices, which are compatible and are adopted in their area.

Before IRM programme, majority of the farmers (41 per cent) reported, only one resistant variety, 21 per cent and five per cent respondents reported, two and three varieties, respectively. After the training 31, 31 and 26 per cent of the respondents reported one, two and three resistant varieties, respectively. The district wise detail of the per cent change in the respondent farmers acquiring knowledge about resistant varieties is given in Table 5.3. The percentage of farmers' having knowledge about Ankur 651 was higher compared to other varieties, it is due to the reason, that the seed company agents of Ankur 651 are popularizing their variety in addition to other Ankur varieties. The influence of private companies is also visible in dissemination of information.

5.4.2 Knowledge About Timely Sowing

Along with cultivation of early maturing resistant varieties, the time of sowing plays an important role in reducing the pest damage and pesticide use. The manipulation of sowing time helps to minimize the build up of *Helicoverpa armigera* (ABW) and *Bemisia tabaci* (whitefly) and timely sowing crop escapes damage from these pests (Dhawan 1999).

A very high percentage of the IRM farmers (81 per cent) were having the knowledge about the recommended month of sowing cotton crop, even before the IRM programme. The percentage of IRM farmers having knowledge about timely sowing (April sowing) increased to 100 per cent after IRM programme. The pre-training knowledge about the benefits of timely sowing with respect to insect pests was also high, as 51 and 13 per cent of the IRM farmers reported that timely sown crop escapes the late damage of ABW and overall pest attack, respectively. In post-training 73 and 20 per cent (overall 93 per cent) of the respondents reported about these two benefits of timely sowing. The added benefit of timely sowing of cotton crop

Table 5.3: Knowledge About Recommended Resistant Varieties

Variety/Reporting About No. of Varieties	District Wise Percentage of Farmers									Overall Percentage of Three Districts (n=150)		
	Bathinda (n=50)			Ferozepur (n=50)			Mansa (n=50)					
	Before IRM	After IRM	Difference	Before IRM	After IRM	Difference	Before IRM	After IRM	Difference	Before IRM	After IRM	Difference
Hybrid varieties												
(i) LHH 144	24	42	18	0	6	6	8	16	8	11	21	10
(ii) Ankur 651[1]	36	68	32	28	46	18	42	72	30	35	62	27
(iii) White Gold[1]	22	30	8	4	14	10	6	20	14	11	21	10
Non-hybrid varieties												
(i) F 1861	22	28	6	16	46	30	0	4	4	13	26	13
(ii) LH 1556	34	40	6	16	48	32	24	34	10	25	41	16
Reporting about												
(i) Three resistant varieties	12	40	28	0	22	22	2	16	14	5	26	21
(ii) Two resistant varieties	38	32	–6	10	32	22	14	28	14	21	31	10
(iii) One resistant variety	32	20	12	44	32	-12	46	40	-6	41	31	–10

[1] Not PAU developed varieties but recommended by PAU

Decimals have been rounded up to nearest whole numbers

Multiple response.

is that the following wheat crop is sown timely, as 29 per cent and 34 per cent reported about this benefit pre and post test, respectively (Table 5.4). The district wise and overall percentage of farmers having knowledge about timely sowing and its benefits are listed in Table 5.4.

5.4.3 Knowledge About Other Cultural Practices

The cultural practices, which play a role in minimizing pest damage to cotton crop, as enlisted in the IRM manual were included in the knowledge test. Though resistant varieties and timely sowing also are part of cultural practices but these were discussed separately. Plant spacing, seed rate, fertilizer and irrigation management, clean cultivation, benefits of early maturing and short duration varieties are also an important aspect of IPM system. The pre and post-training percentage of farmers having knowledge about these cultural practices is given in Table 5.5.

The IRM programme started when the farmers were already in the process of sowing their cotton crop. The pre and post-training percentage of farmers having knowledge about recommended seed rate of hybrid and non-hybrid varieties was low. None of the farmers had pre-training knowledge about recommended seed rate for hybrid cotton varieties and reported less seed rate (900 g/acre), where as PAU recommends 1.5 kg/acre. Even post-training, only three per cent of the farmers reported about correct seed rate. The seed companies recommending 900g/acre is also one of the reasons for this. In case of non-hybrid varieties 13 and 39 per cent of the IRM farmers had knowledge about recommended seed rate, before and after training, respectively (Table 5.5). The PAU has reduced the recommended seed rate in case of non-hybrid varieties of cotton in its *Package of Practices for crops of Punjab kharif–2005*, from 4 kg/acre (10kg ha^{-1}) in 2004 (PAU 2004) to 3.5 kg/acre (8.750 kg ha^{-1}) in 2005 (PAU 2005). During the training programmes much emphasis was not given to seed rate of the different cultivars of cotton.

The percentage of IRM farmers having knowledge about line-to-line spacing was 74 per cent (pre-training) and 93 per cent (post-training), but pre and post-training percentage of IRM farmers having knowledge about plant-to-plant spacing was 31 per cent and 57 per cent, respectively. The proper spacing is desired for reduced pest population build up. IRM farmers had acquired knowledge about manual thinning to maintain proper plant spacing as compared to 52 per cent in pre-training, 82 per cent of the farmers had knowledge about it post-training (Table 5.5).

The percentage of IRM farmers having knowledge about sowing early maturing and short duration varieties was 65 per cent pre-training and 94 per cent post-training.

5.4.3.1 Weed Management

Pre-training, small percentage of farmers had knowledge about different aspects of clean cultivation with reference to build up of different pests. Only four per cent of the respondents had pre-training knowledge about alternate host of spotted bollworm, which changed to 28 per cent, post-training (Table 5.5). Pre-training knowledge about benefits of weed removal with respect to spotted bollworm, whitefly and tobacco caterpillar was low; as only, nine, three and 11 per cent of the IRM farmers, respectively, reported about the benefits of weed removal. There were an appreciable percentage of

Table 5.4: Knowledge About Timely Sowing of Cotton Crop and its Benefits

Timely Sowing/Benefits	*District Wise Percentage of Farmers*									*Overall Percentage of Three Districts (n=150)*		
	Bathinda (n=50)			*Ferozepur (n=50)*			*Mansa (n=50)*					
	Before IRM	*After IRM*	*Difference*	*Before IRM*	*After IRM*	*Difference*	*Before IRM*	*After IRM*	*Difference*	*Before IRM*	*After IRM*	*Difference*
Timely sowing (whole of April)	64	100	36	94	100	6	84	100	16	81	100	19
Benefits of timely sowing#												
(i) Escapes damage due to American bollworm	60	78	18	48	82	34	46	72	26	51	73	22
(ii) Less Pest attack	4	10	6	16	30	14	18	20	2	13	20	7
(iii) Less pesticide applications	12	18	6	20	12	–8	6	8	2	13	13	0
(iv) Higher seed cotton yield	18	18	0	12	22	10	10	20	10	13	20	7
(v) Early and more fruiting	2	2	0	14	12	–2	4	6	2	7	7	0
(vi) Early maturity of cotton crop	16	22	6	10	12	2	24	30	6	17	21	4
(vii) Crop establishment before hot summer	14	12	–2	8	0	–8	6	0	–6	9	5	–4
(viii) Timely wheat sowing	56	66	10	12	14	2	20	22	2	29	34	5

Multiple response.

Decimals have been rounded up to nearest whole number.

Table 5.5: Knowledge About Selected Cultural Practices

Knowledge About	*District Wise Percentage of Farmers*									*Overall Percentage of Three Districts (n=150)*		
	Bathinda (n=50)			*Ferozepur (n=50)*			*Mansa (n=50)*					
	Before IRM	*After IRM*	*Difference*	*Before IRM*	*After IRM*	*Difference*	*Before IRM*	*After IRM*	*Difference*	*Before IRM*	*After IRM*	*Difference*
Spacing												
(i) Line x Line	76	94	18	92	94	2	54	90	36	74	93	19
(ii) Plant x Plant	30	70	40	44	54	10	18	48	30	31	57	26
Seed rate												
(i) Hybrid varieties	0	6	6	0	0	0	0	2	2	0	3	3
(ii) Non–hybrid varieties	26	42	16	8	68	60	6	6	0	13	39	26
Fertilizer dosage												
(i) Urea	0	12	12	2	6	4	0	0	0	1	2	1
(ii) DAP	66	58	– 8	50	42	– 8	56	50	– 6	57	50	– 7
Manual thinning	66	88	22	52	78	26	38	80	42	52	82	30
Alternate host weed of SBW (*Sida* sp., *Abutilon* sp.)	8	36	28	0	24	24	4	24	20	4	28	24
Removing weeds prevents initial build up of												
(i) Spotted bollworm	8	50	42	10	58	48	10	36	26	9	48	39
(ii) Whitefly/CLCuV	4	10	6	2	18	16	4	10	6	3	13	10
(iii) Tobacco caterpillars	26	76	50	2	66	64	4	74	70	11	72	61

Contd...

Table 5.5–Contd...

Knowledge About	*District Wise Percentage of Farmers*									*Overall Percentage of Three Districts (n=150)*		
	Bathinda (n=50)			*Ferozepur (n=50)*			*Mansa (n=50)*					
	Before IRM	*After IRM*	*Difference*	*Before IRM*	*After IRM*	*Difference*	*Before IRM*	*After IRM*	*Difference*	*Before IRM*	*After IRM*	*Difference*
Benefits of sowing early maturing/short duration varieties	80	94	14	68	94	26	48	94	46	65	94	29
Disadvantages of excessive irrigation												
(i) More Bollworm damage												
(ii) More whitefly damage.	36	54	18	20	50	30	14	24	10	23	43	20
(iii) More jassid damage	2	8	6	0	8	8	0	2	2	1	6	5
	8	16	8	12	26	12	12	24	12	11	22	11

Decimals have been rounded up to nearest whole number; DAP: Diammonium phosphate; CLCuV: Cotton leaf curl virus.

respondents having post-training knowledge about the benefits of weed management with respect to spotted bollworm (48 per cent) and tobacco caterpillar (72 per cent). Farmers had real life experience of tobacco caterpillar infestation, as there was out break of this insect pest in 2004.

5.4.3.2 Water Management

The excessive irrigation and late irrigation is conducive for build up of bollworms and whitefly and the percentage of IRM farmer having knowledge about it was 23 and one per cent (pre-training) and 43 and six per cent (post-training), respectively (Table 5.5). The district wise detail is given in Table 5.5.

5.4.3.3 Fertilizer Dosages

The topics of fertilizer dosages were not discussed comprehensively in the meetings. Even IRM printed material given to IRM farmers had no mention of the recommended dosages of different fertilizers to be applied. There was a general recommendation for using the fertilizers as per soil test. The pre and post-training, percentage of farmers having knowledge about dosages of urea for hybrid varieties and non-hybrid varieties was one and two percent, respectively (Table 5.5). Farmers did not possess the knowledge of specific dosages. Their knowledge of dosages was in terms of half bag (25 kg), one bag (50kg), one and a half bags (75kg), and so on. In case of diammonium phosphate, as the recommended dosage is 27kg/acre, which is about half bag; so about 50 per cent farmers had pre and post-training knowledge about it (Table 5.5).

5.4.4 Knowledge About Seed Treatment

The pre and post-training knowledge about seed treating chemicals is reported in Table 5.6. Post-training, there was a increase in percentage of IRM farmers having knowledge about different seed treating chemicals. Even though only five per cent of the farmers referred to Gaucho 70 WS as seed treating chemical, but 47 per cent farmers, mentioned the benefits of treating seed with Gaucho 70 WS. The less percentage of farmers possessing knowledge of seed treating chemicals is due to the reason that seed of hybrid varieties of cotton is sold treated, so farmers did not pay enough attention to this aspect and secondly the training programme started when sowing of cotton crop was already going on.

An overview of the farmers' knowledge about different cultural practices indicates that farmers' had mainly gained knowledge about seed rate of non-hybrid verities of cotton, timely sowing, plant geometry, manual thinning, clean cultivation to avoid pest infestation, seed treating chemicals and benefits of short duration varieties. The trainers need to put greater emphasis on the cultural practices in future training programmes, as in the absence of any effective bio-agents, cultural practices together with use of pesticides according to good agricultural practices are the major components of IPM system in Punjab.

5.4.5 Knowledge About Insecticides

IRM as already discussed in chapter 1 is a component of IPM, and IRM strategy is being disseminated to farmers under broader perspective of IPM. For judicious use of insecticides, the knowledge about different aspects of IRM is must.

Table 5.6: Knowledge About Seed Treating Chemicals and its Benefits

Seed Treating Chemicals	*District Wise Percentage of Farmers*									*Overall Percentage of Three Districts (n=150)*		
	Bathinda (n=50)			*Ferozepur (n=50)*			*Mansa (n=50)*					
	Before IRM	*After IRM*	*Difference*	*Before IRM*	*After IRM*	*Difference*	*Before IRM*	*After IRM*	*Difference*	*Before IRM*	*After IRM*	*Difference*
Seed treating chemicals												
(i) Streptocycline	18	28	10	22	32	10	2	14	12	14	25	11
(ii) Emisan–6	22	34	12	24	36	12	8	44	36	18	38	20
(iii) Gaucho 70 WS	2	2	0	4	14	10	0	0	0	2	5	3
Benefits of seed treatment with Gaucho 70 WS[1]	44	64	20	14	32	18	10	44	34	23	47	24

Decimals rounded up to nearest whole numbers.

[1] It delays jassid attack.

Pre-training, 45, 59, 10, 4, 14 and 7 per cent of the IRM farmers had knowledge about synthetic pyrethroids, organophosphates, organochlorinates, nitroguadine, oxadiazine and naturalyte groups of insecticides, respectively and post-training it was 74, 89, 46, 13, 30 and 21 per cent, respectively (Table 5.7). Pre-training majority of the farmers had the knowledge about synthetic pyrethroids and organophosphates and post-training the percentage of farmers having knowledge about other groups of pesticides had increased. Pre-training 24 per cent of the farmers were not having knowledge about any group of insecticide and for those insecticides were simply an insecticide (poison) meant to kill insect pests. Post-training percentage of farmers reporting three groups of insecticides was 62 per cent. Knowledge about different insecticide groups is required under IRM strategy for avoiding the build up of resistance in *Helicoverpa armigera* (ABW), resurgence of whitefly, and conservation of natural enemies mainly until 90 days after sowing. During training much emphasis was laid on training farmers about rotating the different molecules of insecticides and to avoid spraying one insecticide more than once. District wise percentage of farmers having pre and post-training knowledge about different insecticide groups is enlisted in Table 5.7.

5.4.5.1 Knowledge About Proper Insecticide Use

In the IPM-FFS programmes much attention was not paid to judicious use of pesticides rather no attention was paid to correct pesticide application dosages (van de Fliert 1993, Peshin and Kalra 2000a). Under IRM much emphasis is on appropriate use of pesticides (Russell 2004). The knowledge of farmers about proper use of insecticide was studied as per the four windows of IRM strategy (Refer Chapter 1). The percentage of IRM farmers having pre and post-training knowledge about different aspects of appropriate application of insecticides is given in Table 5.8.

The IRM strategy for conservation of natural enemies is to avoid use of insecticides until 60 days after sowing and to apply endosulfan between 60-90 days after sowing, as it is relatively safe to natural enemies. Percentage of farmers in all the districts having knowledge of not using insecticides until 60 days after sowing was, 89 per cent pre-training and 97 per cent after training, but the knowledge about the rationale behind this was not clear to the farmers, which was explained to them during the course of the IRM programme. The knowledge about the recommended insecticide endosulfan between 60-90 days after sowing, pre-training percentage of farmers having knowledge was 14 per cent and post-training 73 per cent of the farmers had knowledge about it. Pre-training, 60 per cent of the IRM farmers lacked knowledge about avoiding monocrotophos up to 90 days after sowing, and reported monocrotophos as the insecticide to be used between 60-90 days after sowing, and post-training a small percentage of five per cent of farmers reported about this insecticide, this shows that majority of the farmers had gained knowledge that broad spectrum insecticide like monocrotophos should not be applied initially. The farmers mostly used monocrotophos for better vegetative growth and not for pest management. The district wise pre and post-training percentage of farmers having knowledge about these aspects is given in Table 5.8. Under the IRM strategy, the use of synthetic pyrethroids is to be avoided during boll formation phase as it leads to its resurgence.

Table 5.7: Knowledge About Different Groups of Insecticides

Group of Insecticide/ No. of Groups	District Wise Percentage of Farmers									Overall Percentage of Three Districts (n=150)		
	Bathinda (n=50)			Ferozepur (n=50)			Mansa (n=50)					
	Before IRM	After IRM	Difference	Before IRM	After IRM	Difference	Before IRM	After IRM	Difference	Before IRM	After IRM	Difference
Synthetic pyrethroids	56	88	32	56	76	20	24	58	34	45	74	29
Organophosphates	86	98	12	68	94	26	24	74	50	59	89	30
Organochlorinates	14	32	18	12	66	54	4	40	36	10	46	36
Nitroguadine	8	8	0	4	8	4	0	24	24	4	13	9
Oxadiazine (Indoxacarb 15 SC)	26	38	12	8	26	18	8	26	18	14	30	16
Naturalyte (Spinosad 48 SC)	12	30	18	4	20	16	6	12	6	7	21	14
Reporting about insecticides belonging to:												
(i) Four groups	4	18	14	2	16	14	0	12	12	2	15	13
(ii) Three groups	38	72	34	8	68	60	12	44	32	19	62	43
(iii) Two groups	28	6	–22	52	12	–40	22	24	2	34	14	–20
(iv) One group	20	4	–16	20	24	–16	22	20	–2	21	9	–12
(v) None of the group	10	0	–10	18	0	–18	44	0	–44	24	0	–24

Decimals have been rounded up to nearest whole numbers.

Table 5.8: Knowledge About Proper Insecticide Use Under IRM Strategy Against Different Insect-pests

Strategy	*District-wise Percentage of Farmers*									*Overall Percentage of Three Districts (n=150)*		
	Bathinda (n=50)			*Ferozepur (n=50)*			*Mansa (n=50)*					
	Before IRM	*After IRM*	*Difference*	*Before IRM*	*After IRM*	*Difference*	*Before IRM*	*After IRM*	*Difference*	*Before IRM*	*After IRM*	*Difference*
No insecticide application upto 60 DAS	92	94	2	92	100	8	84	96	12	89	97	8
Insecticide recommended during 60-90 DAS												
(i) Endosulfan 35 EC*	12	74	62	24	86	62	6	58	52	14	73	59
(ii) Monocrotophos 36 SL** (Lack of knowledge)	48	2	46	60	2	58	72	10	62	60	5	55
Knowledge about												
(i) Resurgence of whitefly by using synthetic pyrethroids	82	94	12	64	82	18	14	72	58	53	83	30
(ii) Insecticide for whitefly during boll formation phase	32	64	32	12	46	34	4	20	16	16	43	27
Knowledge about recommended insecticides against different stages of American bollworm larvae												
(i) Young larvae	54	92	38	24	82	58	20	56	36	33	77	44
(ii) Grown up larvae	46	90	44	22	72	50	18	52	34	29	71	42
(iii) In case of out–break	78	96	18	36	94	58	46	64	18	53	85	32

Contd...

Table 5.8–Contd...

Strategy	*District-wise Percentage of Farmers*									*Overall Percentage of Three Districts (n=150)*		
	Bathinda (n=50)			*Ferozepur (n=50)*			*Mansa (n=50)*					
	Before IRM	*After IRM*	*Difference*	*Before IRM*	*After IRM*	*Difference*	*Before IRM*	*After IRM*	*Difference*	*Before IRM*	*After IRM*	*Difference*
Non effective insecticides against American bollworm												
(i) Monocrotophos 36 SL	16	62	46	18	30	12	2	18	16	12	37	25
(ii) Synthetic pyrethroids	2	18	16	18	32	14	4	8	4	8	19	11
No tank mixing of insecticides	16	96	80	4	92	88	0	86	86	7	91	84
Rotation of insecticide molecules	86	100	14	72	98	26	32	78	46	63	92	29
Spray technique	94	100	6	96	100	4	94	100	6	95	100	5

DAS: Days after sowing; #: Recommended under IRM as it is relatively safe to natural enemies,
##: Not recommended as it is highly toxic to natural enemies; pre-training the percentage of farmers reporting about monocrotophos was high.

About 53 per cent of IRM farmers were having pre-training knowledge that synthetic pyrethroids should not be used as it increases whitefly infestation and post-training 83 per cent, of the IRM farmers were having knowledge about this fact. But the majority of the farmers were not having knowledge about the insecticides recommended during boll formation phase. Pre-training, 16 per cent, and post-training, 43 per cent of the IRM farmers had knowledge about the recommended insecticides *viz.* triazophos and ethion. Majority of the farmers reported acetamiprid as the recommended insecticide for whitefly during boll formation phase. In district Bathinda, the scenario was different as post-training 64 per cent of the IRM farmers had knowledge about the recommended insecticides for whitefly during the boll formation phase (Table 5.8).

The percentage of farmers having pre-training knowledge about the recommended insecticides for different stages of American bollworm was ranging between 29 to 53 per cent. The percentage of farmers having knowledge that profenophos/triazophos/ethion are the recommended insecticide for young larvae of bollworms was 33 per cent (pre-training) and 77 per cent (post-training). In case of older larvae of bollworms 29 per cent of the respondents had pre-training knowledge and 71 per cent had post-training knowledge. In case of outbreak of American bollworm, 53 per cent of IRM farmers had pre-training knowledge, and 85 per cent had post-training knowledge regarding the recommended insecticides. Pre-training knowledge was due to complimentary factor of pesticide companies, as during 2003 season new molecules of indoxacarb and spinosad were used for American bollworm, and the results were observable to farmers. As a result more than 50 per cent of the respondents had pre-training knowledge about these chemicals and in Bathinda district it was as high as 78 per cent (Table 5.8).

American bollworm has developed resistance against certain insecticides, especially synthetic pyrethroids but the farmers' knowledge about it was low. After training, 19 per cent were having knowledge about synthetic pyrethroids being non-effective against American bollworm. The percentage of farmers having knowledge about not tank mixing insecticide was seven (pre-training), which increased to 91 per cent (post-training).

Under IRM strategy insecticide are to be rotated during a cropping season to avoid development of resistance. Sixty three per cent of the farmers, had knowledge that insecticides should be rotated, that no one insecticide should be used again and again. Post-training the percentage of farmers having knowledge about this aspect was 92 per cent. The farmers had pre-training knowledge about the spray technique as 95 per cent reported that the droplet size of the spray should be as small as possible and it should form a mist. Farmers were not seen using protective clothes while spraying insecticides (Figure 5.11), and no attention was paid to human aspects regarding the adverse effects of insecticides on human health during the training. It is a sad state of affairs that protection from pesticide exposure is poor.

Overall, farmers have gained knowledge about different aspects of proper insecticide use, but the knowledge gain about insecticide recommended for whitefly during boll formation phase, not to use synthetic pyrethroids against American

(a)

(b)

Figure 5.11: Health Hazards: Farmers Applying Pesticides without Protective Clothing (a) Manual sprayer; (b) Tractor mounted sprayer

bollworm was low. The district wise detail about percentage of farmers having knowledge about different aspects of proper insecticide use is given in Table 5.8.

5.4.5.2 Reasons for Reduction of Pesticide Efficacy

The efficacy of pesticides having reduced over a period of time is known throughout the different agricultural systems. The farmers are also aware of this fact, as pesticides do not control the pest problem. Pre-training 63 per cent, of the IRM farmers were aware that pests have developed resistance to insecticides which they termed '*Amli*' in local language meaning 'used to', but 37 per cent of the IRM farmers said that they were not aware of any such thing and there is nothing called resistance, but it is mainly due to poor quality of insecticides. Post-training 89 per cent, of the IRM farmers were aware about the resistance syndrome (Table 5.9).

The reasons given by farmers about reduction of pesticide efficacy were many. From resistance in insect pests, repeated use/excessive use of pesticides, indiscriminate dosage, poor quality of pesticides to climate change, destruction of natural enemies, spray technology, more pest infestation, etc. Few farmers reported that due to greediness on the part of farmers and pesticide dealers, the situation has reached to such an alarming stage. The detail about different reasons given by farmers for reduction of pesticide efficacy is given in Table 5.9.

5.4.6 Knowledge About Insect Pests

IRM farmers knowledge about major insect pests of cotton crop was tested in the base line survey (before training) and at the end of IRM programme in 2004 (after training). The IRM farmers were asked to identify the insect pests from real specimen and damage symptoms from photographs. Figure 5.12 shows the change in cognition of IRM farmers before and after training regarding different insect pests.

Before training, IRM farmers already possessed high knowledge about jassid, whitefly and American bollworm. The percentage of IRM farmers having knowledge about jassid, second injury grade of jassid, whitefly and American bollworm was 83, 75, 83 and 89 per cent, respectively. In district Bathinda, pre-training knowledge about these insect pests was higher in comparison to other districts (Table 5.10). Some of the villages in Bathinda (Malwala) and Ferozepur (Wahabwala) were previously covered under IPM-FFS in late 1990s by the State Agricultural Department and central integrated pest management center of Directorate of Plant Protection, Quarantine and Storage, Government of India.

The knowledge about other insect pests like spotted bollworm, aphid and tobacco caterpillar was low in pre-training identification of these insect pests (Table 5.10). During pre-training identification majority of the IRM farmers identified spotted bollworm wrongly as American bollworm and could not differentiate the damage symptoms of these two insect pests. During 2004, season spotted bollworm infestation was more compared to previous season, so the farmers experienced the damage symptoms of this insect pest which were shown to them by the trainers in the field. The knowledge gain was very high. Pre-training only 31 per cent of IRM farmers identified the spotted bollworm and post-training 73 per cent identified the same insect pest. This shows that 42 per cent of the trainees gained knowledge about

Table 5.9: Knowledge and Reasons for Reduction of Pesticides Efficacy Over a Period of Time

Knowledge About/Reason	*District Wise Percentage of Farmers*									*Overall Percentage of Three Districts (n=150)*		
	Bathinda (n=50)			*Ferozepur (n=50)*			*Mansa (n=50)*					
	Before IRM	*After IRM*	*Difference*	*Before IRM*	*After IRM*	*Difference*	*Before IRM*	*After IRM*	*Difference*	*Before IRM*	*After IRM*	*Difference*
Insecticide resistance in insect pests												
Aware	80	96	16	62	84	22	46	86	40	63	89	26
Reasons for reduction in efficacy of insecticides*												
(i) Resistance in insect pest	46	58	12	30	78	48	22	36	14	33	57	24
(ii) Poor quality insecticides	42	30	–12	56	56	0	70	56	–14	56	47	–9
(iii) Excessive use of insecticides	46	64	18	22	22	0	10	20	10	26	36	10
(iv) Indiscriminate/less dosage of insecticides	22	44	22	12	12	0	4	6	2	13	21	8
(v) Repeated use of same insecticides	6	20	14	0	0	0	0	4	4	2	8	6
(vi) Tank mixing of insecticides	4	24	20	2	4	2	2	12	10	3	13	10
(vii) Climate change	4	6	2	6	2	–4	2	2	0	4	3	–1
(viii) Spray technology	2	2	0	0	0	0	2	2	0	1	1	0
(ix) Destruction of natural enemies	2	2	0	2	2	0	0	0	0	1	1	0
(x) No proper dilution of insecticides with water	2	2	0	0	0	0	0	0	0	1	1	0
(xi) Greediness of farmers and dealers	0	0	0	4	4	0	0	0	0	1	1	0
(xii) More pest (ABW)	2	2	0	2	0	–2	6	8	2	3	3	0

Decimals rounded up to nearest whole numbers.

Per cent difference showing positive change after IRM programme.

Table 5.10: Knowledge About Major Insect Pests and Natural Enemies of Cotton Ecosystem

Insect Pest/Natural Enemy Identified	*District Wise Percentage of Farmers*									*Overall Percentage of Three Districts (n=150)*		
	Bathinda (n=50)			*Ferozepur (n=50)*			*Mansa (n=50)*					
	Before IRM	*After IRM*	*Difference*	*Before IRM*	*After IRM*	*Difference*	*Before IRM*	*After IRM*	*Difference*	*Before IRM*	*After IRM*	*Difference*
Insect pests												
(i) Jassid	94	98	4	88	94	6	66	88	22	83	93	10
(ii) Second injury grade of jassid	86	98	12	68	96	28	72	94	22	75	96	21
(iii) Whitefly	90	98	8	90	96	6	68	94	26	83	96	13
(iv) American bollworm	94	100	6	90	100	10	84	100	16	89	100	11
(v) Spotted bollworm	34	82	48	38	74	36	22	64	42	31	73	42
(vi) Aphid	26	48	22	22	44	22	8	16	8	19	36	17
(vii) Tobacco caterpillar	36	100	64	12	86	74	10	86	76	19	91	72
Natural enemy												
I. *Identified as natural enemy*												
(i) Spiders	86	98	12	66	88	22	76	98	22	76	95	19
(ii) Lady bird beetle	38	86	48	30	76	46	30	74	44	33	79	46
(iii) Predatory bugs	0	8	8	6	28	22	2	4	2	3	13	10
(iv) Green lacewing	16	48	32	16	52	36	6	44	38	13	48	35

Contd...

Table 5.10–Contd...

Insect Pest/Natural Enemy Identified	*District Wise Percentage of Farmers*									*Overall Percentage of Three Districts (n=150)*		
	Bathinda (n=50)			*Ferozepur (n=50)*			*Mansa (n=50)*					
	Before IRM	*After IRM*	*Difference*	*Before IRM*	*After IRM*	*Difference*	*Before IRM*	*After IRM*	*Difference*	*Before IRM*	*After IRM*	*Difference*
II. *Identified but not knowing if natural enemy*[1]												
(i) Spiders	4	0	4#	12	2	10#	4	0	4#	7	1	6#
(ii) Lady bird beetle	18	0	18#	34	2	32#	16	2	14#	23	1	22#
(iii) Predatory bugs	0	0	0	4	4	0	0	0	0	1	1	0
(iv) Green lacewing	2	4	–2	10	4	6#	6	0	6#	6	3	3#
III. *Identified as insect pest*[1]												
(i) Spiders	4	0	4#	0	0	0	0	0	0	1	0	1#
(ii) Lady bird beetle	12	6	6#	4	0	4#	0	0	0	5	2	3#
(iii) Green lacewing	8	2	6#	2	0	2#	0	0	0	3	1	2#

Decimals rounded up to nearest whole numbers, [1] Lack of knowledge.

#: Per cent difference showing positive change after IRM programme.

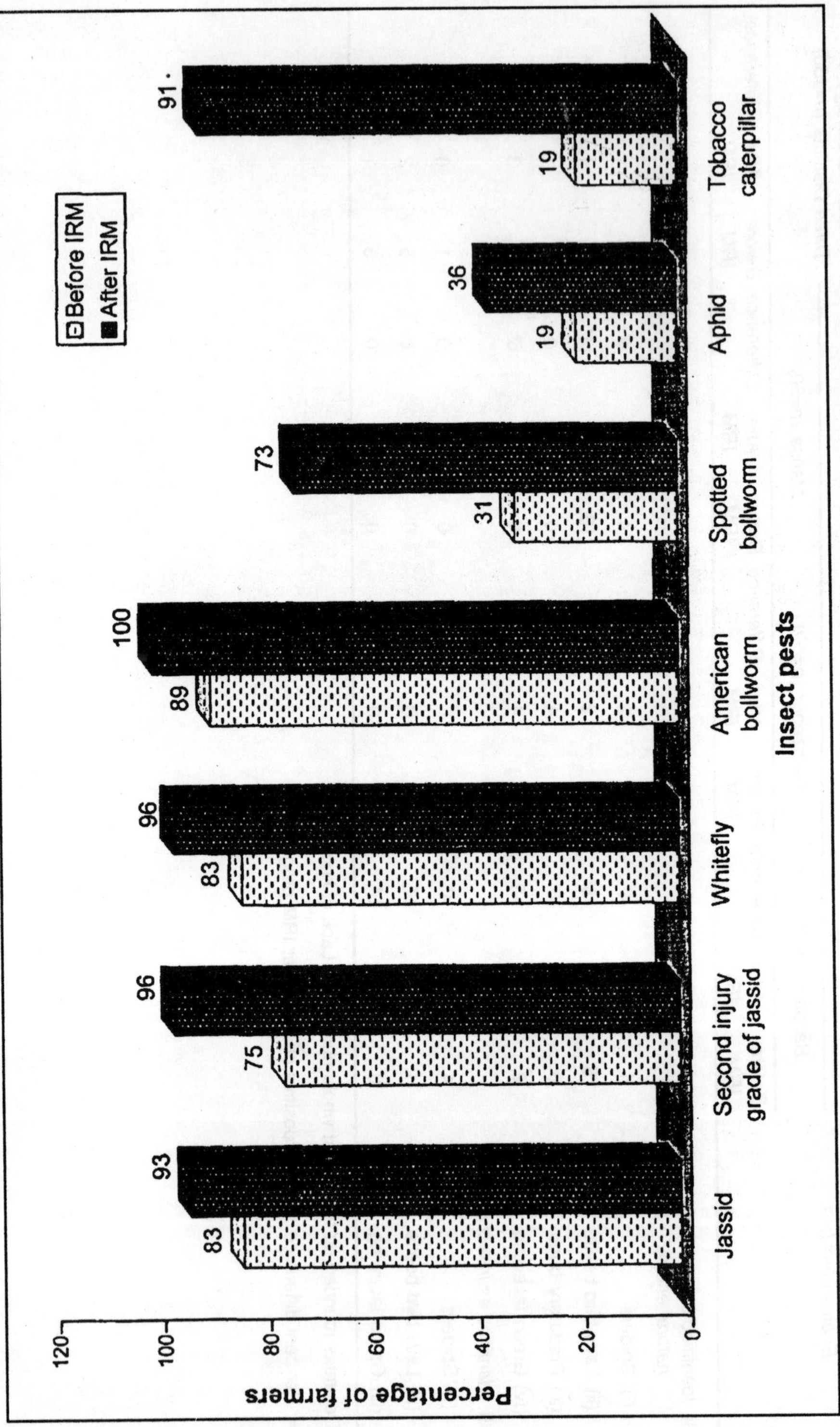

Figure 5.12: Knowledge About Insect Pests Before and After IRM Programme

spotted bollworm. Tobacco caterpillar was barely known pest before the IRM programme as only 19 per cent, of the respondents identified this pest. During 2004, season there was severe outbreak of this pest in all the three districts of Punjab. The knowledge gain about this insect pest was highest (72 per cent) as post-training 91 per cent of the respondents identified this insect pest. The farmers discovered the mode of damage and spread of this insect pest, and due to IRM programme got timely advice for manual mechanical and chemical control of this insect pest. The pre-training and post-training knowledge about aphid identification was low. Overall the IRM farmers of district Bathinda were more knowledgeable about insect pests.

5.4.7 Identification of Natural Enemies

Recognition of natural enemies and knowledge about their preys is the first competency objective relating to IPM principle on natural enemy conservation (van de Fliert 1993). The pre-training IRM farmers' cognition about natural enemies in cotton ecosystem was measured by using specimen and photographs of natural enemies. A very high percentage of farmers, 76 per cent, identified spiders and as knew that it preys on insect pests. But the percentage of farmers having knowledge about lady bird beetle being a natural enemy and preying on jassid and other insect pests was moderate (33 per cent). The farmers reported that some 10 years back this natural enemy was observed in abundance, but now rarely seen in the cotton fields. The percentage of farmers having pre-training knowledge about predatory bugs and green lacewing was three and 13 per cent, respectively (Table 5.10). Post-training knowledge of spiders increased from 76 to 95 per cent, lady bird beetle, 33 to 79 per cent, green lacewing 13 to 48 per cent and predatory bugs 3 to 13 per cent (see Figure 5.13). The reason for low increase in the percentage of farmers having knowledge about predatory bugs was that it was not as abundant as other natural enemies in the cotton fields (Dhawan 2005).

Pre-training, some percentage of the farmers even though identified natural enemies but reported these either to be insect pests or showed ignorance about these being a natural enemy. The IRM farmers gained knowledge about the role of these natural enemies during the IRM programme. Before training; seven, twenty-three, one and six per cent identified spiders, lady bird beetle, predatory bugs and green lacewing, respectively, but were not knowing if these are predators of insect pests. Post-training, the lack of knowledge about this aspect was one to three per cent in case of these natural enemies. The incorrect knowledge about spiders, lady bird beetle and green lacewing being pests was also corrected during the IRM programme. Pre-training, one to 5 per cent of farmers reported spiders; green lacewing and lady bird beetle as pests and post-training hardly one to two per cent reported green lacewing and lady bird beetle as pests (Table 5.10).

The IRM farmers were trained about conservation of natural enemies in cotton crop until 90 days after sowing. The special topics on–"not to apply insecticides until 60 days after sowing, and effect of monocrotophos on natural enemies" were discussed. The literature regarding these aspects was distributed in the form of leaflets, etc. The farmers were told to avoid insecticides until 90 days after sowing, and in case of emergency, use endosulfan which is relatively safe to natural enemies. The gain in

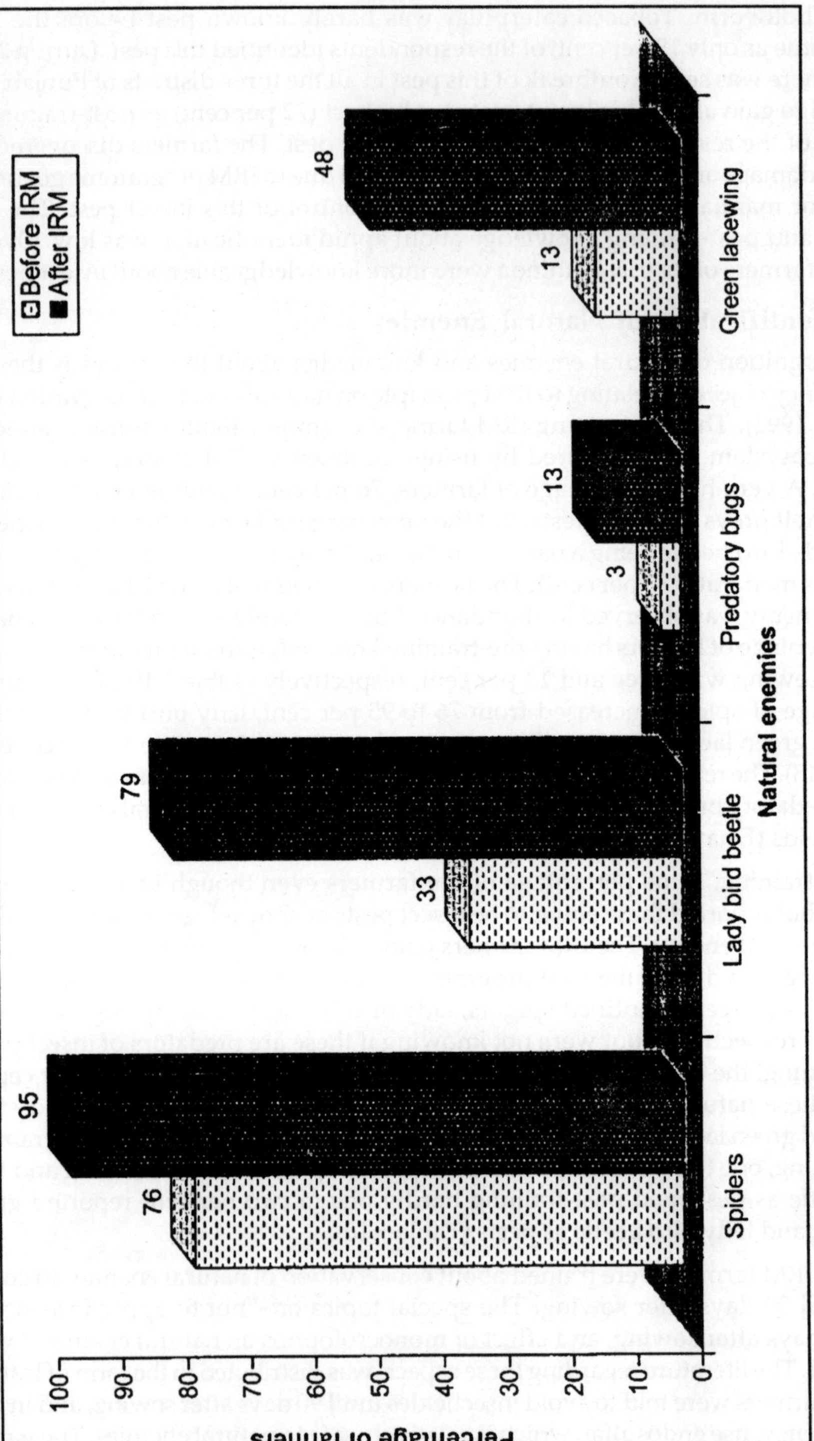

Figure 5.13: Knowledge About Natural Enemies Before and After IRM Programme

knowledge resulted in the reduction in the use of monocrotophos in cotton crop and farmers on an average applied first insecticide application after 86. 49 days after sowing (refer Table 5.32).

5.4.8 Knowledge About Economic Threshold Level

The PAU, Ludhiana recommended the ETL of jassid in 1979 in its package of practices (PAU 1979) and for whitefly and bollworms in 1991 (PAU 1991). But it was observed that farmers in IRM villages had almost no knowledge about ETL (Table 5.11). Pre-training, the percentage of farmers having knowledge about ETL of jassid, whitefly and bollworms was three, one and one per cent, respectively, and post-training for the same pests, the percentage of IRM farmers having knowledge was 42, 37 and 41 per cent, respectively (see Figure 5.14). The dissemination of complex technologies like ETL or even IPM as such through traditional mode of transfer of technology (ToT) is not going to spread the information, knowledge and adoption of these practices. This has again proved the hypothesis that ToT model was appropriate for the green revolution technologies and not for knowledge intensive technologies like IPM.

5.4.9 Overall Knowledge Gain of Farmers

Simple statistical tests are necessary to give an overview to the data and to summarize the results based on statistical inferences. The overall knowledge gain of IRM farmers about different practices of IPM was significant at $p < 0.01$. Based on Singh (1975) cube root method, the knowledge scores of the farmers were categorized into three categories of low, medium, high (Table 5.12).

The IRM farmers were having low and medium knowledge level before training. Fifty per cent, of the farmers had low knowledge level, 48 per cent had medium knowledge level and only two per cent had high knowledge level. Post IRM programme, the trend had almost reversed, as four per cent had low knowledge level, 31 per cent medium knowledge level and 65 per cent high knowledge level (Figure 5.15). In district Bathinda, majority of the respondents (74 per cent) had high pre-training knowledge level, whereas in Mansa, majority of the respondents (76 per cent) had low pre-training knowledge level, and in Ferozepur about fifty-fifty per cent had low (54 per cent) and medium (46 per cent) pre-training knowledge level. Post-training, majority of the IRM farmers possessed high knowledge level in Bathinda (84 per cent), Ferozepur (66 per cent), whereas in Mansa they moved from low knowledge level to medium knowledge level (46 per cent) and high knowledge level (46 per cent). Overall 65 per cent, of the IRM farmers possessed high knowledge level after training (Table 5.12).

The mean knowledge score before training was 24.66 (Bathinda), 19.59 (Ferozepur) and 15.81 (Mansa) which was statistically having significant difference at $p<0.05$. Post-training, these mean scores improved to 38.22 (Bathinda), 35.11 (Ferozpur) and 30.69 (Mansa). The difference was statistically significant between the districts of Bathinda and Ferozepur ($p<0.05$), Bathinda and Mansa ($p<0.01$) and Ferozepur and Mansa ($p<0.01$). Knowledge gain of farmers was reported by Rossing *et al.* (1985), Zadoks (1989a), van de Fliert (1993) and others. For IPM and its

Table 5.11: Knowledge About Economic Threshold Level of Different Cotton Insect Pests

ETL of Insect Pest	*District-wise Percentage of Farmers*									*Overall Percentage of Three Districts (n=150)*		
	Bathinda (n=50)			*Ferozepur (n=50)*			*Mansa (n=50)*					
	Before IRM	*After IRM*	*Difference*	*Before IRM*	*After IRM*	*Difference*	*Before IRM*	*After IRM*	*Difference*	*Before IRM*	*After IRM*	*Difference*
Jassid (Curling along leaf margins on 50 per cent plants)	6	50	44	2	48	46	2	28	26	3	42	39
Whitefly (6 adults per leaf)	4	54	50	0	42	42	0	16	16	1	37	36
Bollworm complex (5 per cent incidence in freshly shed fruiting bodies)	2	50	48	2	42	40	0	30	30	1	41	40
Technique of finding ETL of bollworm complex	6	56	50	2	54	52	0	42	42	3	51	48

Decimals rounded up to nearest whole numbers.

Table 5.12: Overall Knowledge Gain and Mean Knowledge Score of Farmers About IPM Practices Disseminated Under IRM Programme

Knowledge Level/Score	*District-wise Percentage of Farmers*									*Overall Percentage of Three Districts (n=150)*		
	Bathinda (n=50)			*Ferozepur (n=50)*			*Mansa (n=50)*					
	Before IRM	*After IRM*	*Difference*	*Before IRM*	*After IRM*	*Difference*	*Before IRM*	*After IRM*	*Difference*	*Before IRM*	*After IRM*	*Difference*
Knowledge level												
(i) Low (0–19.31)	10 (20)	0	–10 (–20)	27 (54)	2 (4)	–25 (–50)	38 (76)	4 (8)	–34 (–68)	75 (50)	6 (4)	–69 (–46)
(ii) Medium (19.31– 33.37)	37 (74)	8 (16)	–29 (–58)	23 (46)	15 (30)	–8 (–16)	12 (24)	23 (46)	+11 (22)	72 (48)	46 (31)	–26 (–17)
(iii) High (33.37–52)	3 (6)	42 (84)	+39 (78)	0	33 (66)	+33 (66)	0	23 (46)	+23 (46)	3 (2)	98 (65)	+95 (63)
Mean knowledge score	24.66	38.22	+13.56*	19.59	35.11	+15.16*	15.81	30.69	+14.88*	20.17	34.70	+14.53*
SD	5.58	6.12	5.06	5.43	5.37	5.85	5.66	7.41	6.01	6.76	7.46	5.66

Figures in the parenthesis are percentages.

Decimals have been rounded up to nearest whole numbers in case of per cent farmers.

*: The paired difference significant at P>0.01 as calculated 't' values 19.143, 18.491, 17.508 and 31.651 in case of Bathinda, Ferozepur, Mansa and overall for three districts, respectively are higher than table values.

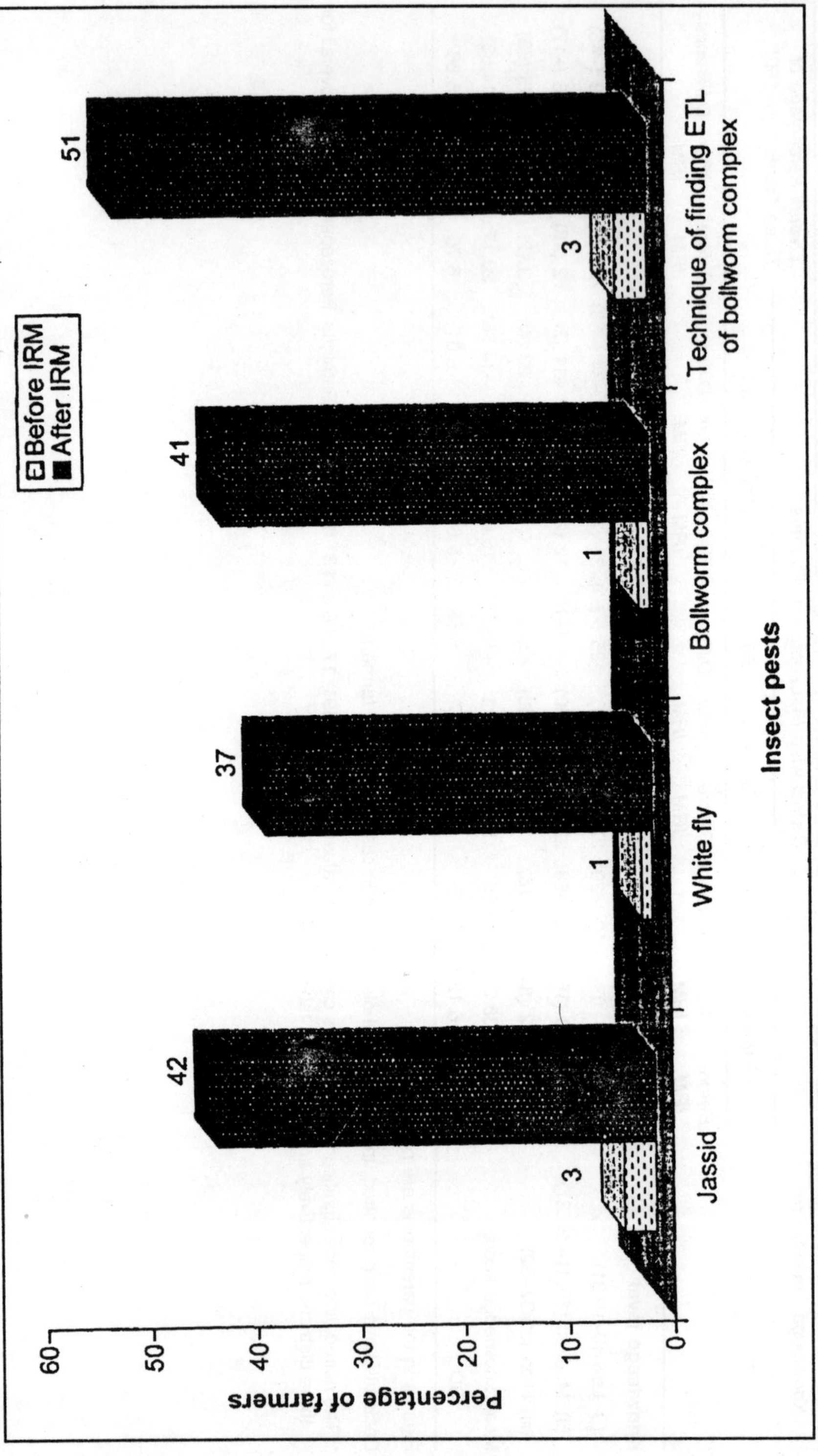

Figure 5.14: Knowledge About ETL of Insect Pests Before and After IRM Programme

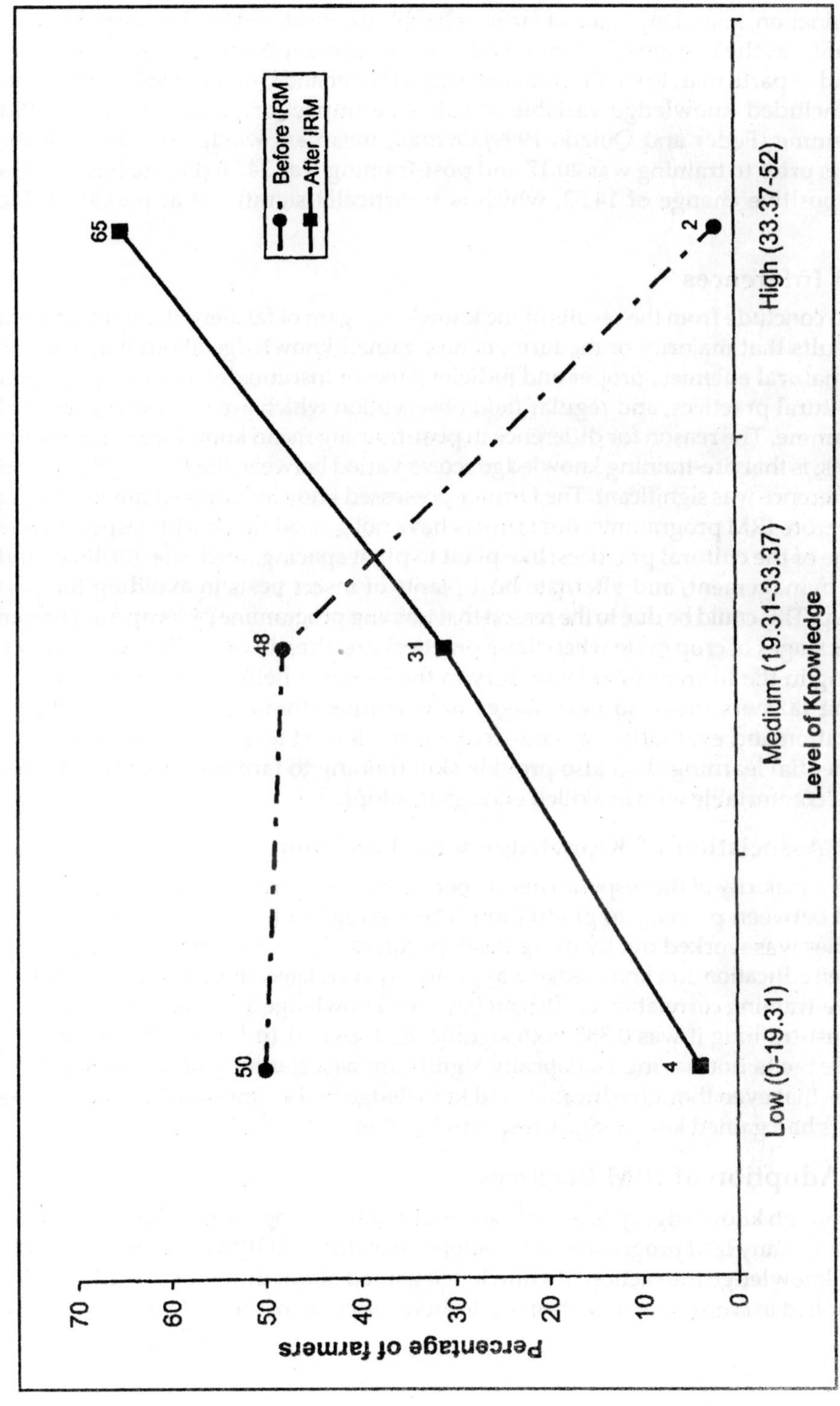

Figure 5.15: Overall Knowledge Gains of Farmers About IPM Practices

dissemination, knowledge gain of farmers has been considered for the desired impact on profits, as the rise in knowledge leads to change in the input mix and in practice use, and in particular, to small use of pesticides (Feder and Quizon 1999). Economists have included knowledge variable to calculate impact or gross benefits of IPM programme (Feder and Quizon 1999).Overall, mean knowledge score for all the districts prior to training was 20.17 and post-training was 34.70 (Figure 5.16) with a mean positive change of 14.53, which is statistically significant at $p<0.01$ (Table 5.12).

5.4.10 Inferences

To conclude from the results of the knowledge gain of farmers, it is evident from the results that majority of the farmers had gained knowledge about major insect pests, natural enemies, proper and judicious use of insecticides according to good agricultural practices, and regular field observation which are the rewards of IRM programme. The reason for difference in post-training mean knowledge score of the districts, is that pre-training knowledge scores varied between the three districts and the difference was significant. The farmers possessed knowledge about timely sowing even before IRM programme. But farmers have not gained much with respect to role of some of the cultural practices; like plant to plant spacing, seed rate, fertilizer and water management, and alternate host plants of insect pests in avoiding the pest build up. This could be due to the reason that training programme picks up momentum in later stages of crop cycle when these practices are already over. The experimental learning in the form of field laboratory in the farmer's field will go a long way in making farmers move to next stages of cognitive domain *i.e.* comprehension, application and evaluation, as knowledge is the lowest level of cognitive domain. Experiential learning shall also provide skill training to farmers about which they are not comfortable with or skilled enough to adopt.

5.4.11 Association of Knowledge with Education

The majority of the respondents (90 per cent) were literate, but level of education varied between primary to graduation. The association if any, between these two variables was worked out by using Pearson correlation coefficient. The association between education and knowledge was positively correlated and significant at $p<0.01$. The pre-training correlation coefficient between knowledge and education was 0.343 and post-training it was 0.383 both significant at $p<0.01$ (d.f.=148). But the gain in knowledge is not having statistically significant association with education. This implies that even though education and knowledge had a significant association but farmers had gained knowledge irrespective of their level of education.

5.5 Adoption of IPM Practices

Though knowledge gain is the first reward of IPM programmes, but the ultimate objective of any IPM programme is the adoption of different IPM tactics which may be called knowledge into action. If knowledge gained about different IPM practices is not applied in crop cultivation then the desired objectives of the IPM are not achieved.

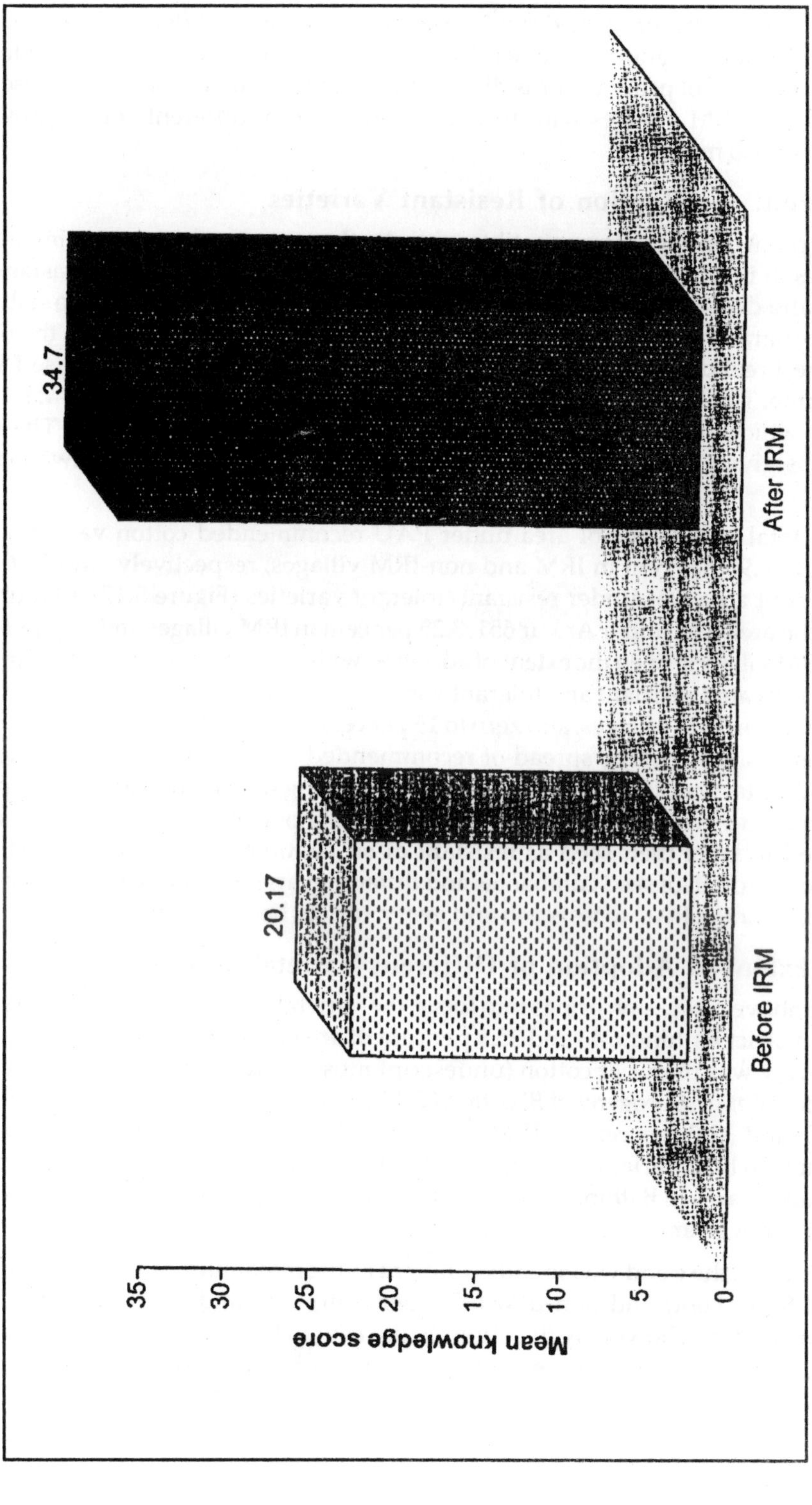

Figure 5.16: Mean Knowledge Score About IPM Practices Before and After IRM Programme

Even though, the farmers gained significant knowledge about different IPM practices during the IRM programme, whether they applied acquired knowledge at their farms during the course of programme is discussed under this sub section. Comparison of IRM and non-IRM villages with respect to adoption of different cotton growing practices is also made.

5.5.1 Extent of Adoption of Resistant Varieties

The plant resistance is compatible with all other management tactics, including pesticides. It is ideally suited under IPM system. As long as host plant resistance is reliable, and does not quickly select for resistance-breaking pests, it helps to stabilize on IPM programme that uses host plant resistance (Norris *et al.*, 2002). On the other hand, the breakdown in resistance can destabilize, plant resistance based IPM programme. The PAU has recommended Ankur 651, White gold, LHH 144, F 1378, LH 1556 which are resistant/tolerant to CLCuV and jassid (PAU 2004). The IRM guidelines given to farmers also enlisted these resistant varieties (Dhawan *et al.*, 2004).

The total percentage of area under PAU recommended cotton varieties was 12.92 and 16.95 per cent in IRM and non-IRM villages, respectively, and 7.16 and 6.22 per cent area was under resistant/tolerant varieties (Figure 5.17). Out of this most of the area was under Ankur 651, 3.25 per cent in IRM villages and 3.65 per cent in non-IRM villages. Even the extent of adoption with respect to percentage of farmers having cultivated resistant and tolerant varieties was ranging between one per cent to 17 per cent in IRM villages, and zero to 15 per cent in non-IRM villages (Table 5.13). This goes to show that the spread of recommended resistant varieties is limited in Punjab. This tactics of IPM is not significant for taking it as an indicator of adoption of IPM practices in cotton crop. The seed companies should be asked by the authorities concerned to label their seed as resistant or susceptible to a particular pest. The mushroom growth of seed companies and different varieties makes it difficult rather impossible for the PAU to test each and every variety being sold in Punjab.

5.5.1.2 Extent of Adoption of Different Varietal Groups

The above discussion brings into focus, the varietal pattern followed in the study area. In district Bathinda, 43 and 35.9 per cent of area in IRM and non-IRM villages, respectively was under *Bt* cotton (undescript mostly) and 17.1 and 11.6 per cent of area was under first progeny of *Bt* cotton (Table 5.14). The area under hybrid varieties was 27.7 and 38.0 per cent in IRM and non-IRM villages, respectively, and area under non-hybrid varieties was 9.7 and 12.4 per cent. This shows that maximum percentage of area in Bathinda was under Bt cotton, even though prior to 2005, it had not got the government approval for cultivation in Punjab.

In Mansa, the trend was slightly different as the maximum percentage of area in IRM (61.8 per cent) and non-IRM (45.5 per cent) was under hybrid varieties of American cotton followed by Bt cotton, 20 per cent in IRM and 24.1 per cent in non-IRM villages. The second progeny of Bt cotton covered an area of 8.3 per cent (IRM villages) and 5.4 per cent (non-IRM villages).

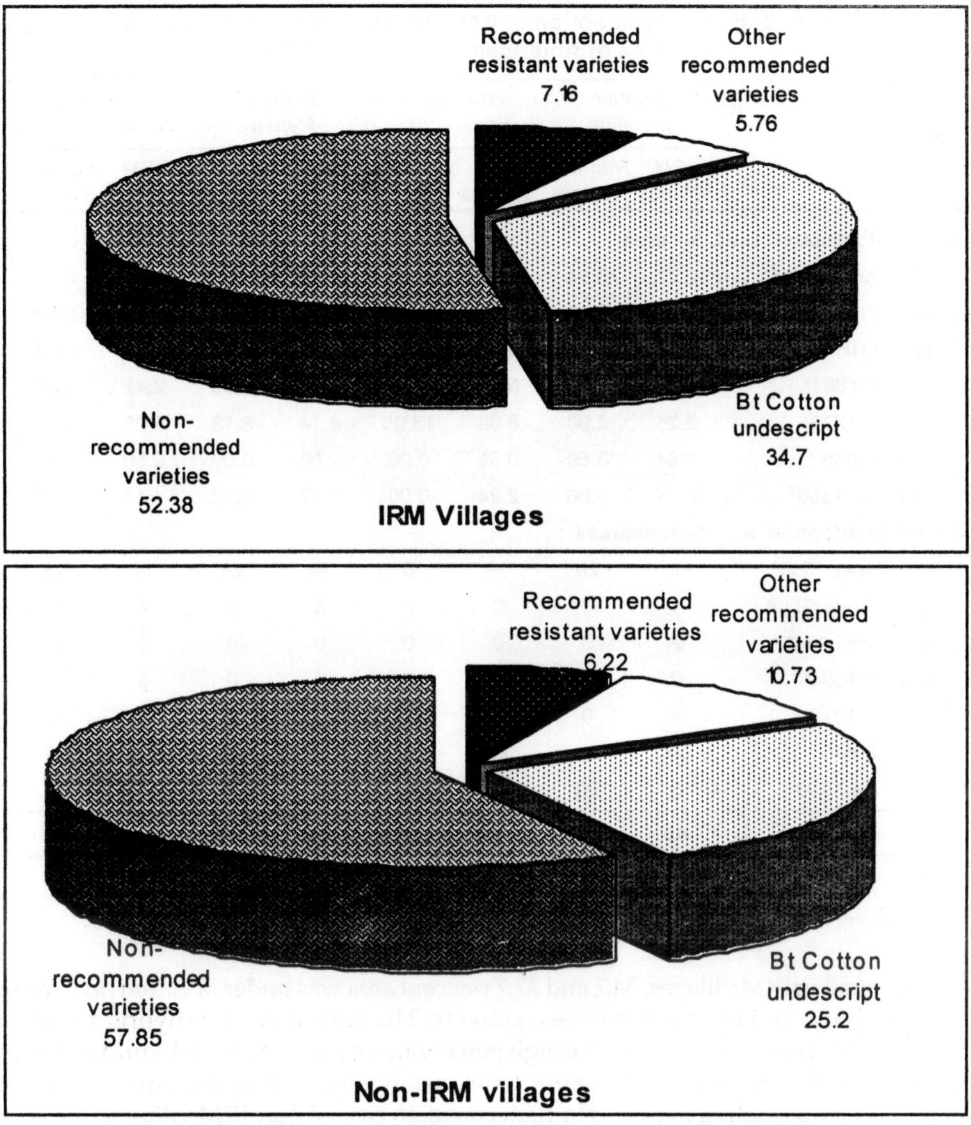

Figure 5.17: Extent of Area Under Different Varieties of Cotton

In case of Ferozepur (Abohar) the trend was completely different. The maximum area was under non-hybrid varieties of American cotton in both IRM villages (80.9 per cent) and non-IRM villages (77.6 per cent), followed by hybrid/Bt cotton (Table 5.14). The overall area under *desi* cotton was 1.1 per cent in IRM area and 1.5 per cent in non-IRM area of all the three districts. The overall varietal pattern is given in Table 5.14.

Table 5.13: Extent of Adoption of PAU Recommended Resistant/Other Varieties of American Cotton During 2004

	Variety	District Bathinda		District Ferozepur		District Mansa		Overall of Three Districts	
		IRM Villages	Non-IRM Villages	IRM Villages	Non-IRM Villages	IRM Villages	Non-IRM Villages	IRM Villages	Non-IRM Villages
Extent of adoption w.r.t. % area									
(i)	Ankur 651#	2.77	5.54	1.68	0.00	7.90	12.95	3.25	3.65
(ii)	White Gold#	0.42	0.00	0.00	0.00	0.92	0.00	0.34	0.00
(iii)	LHH 144#	0.06	0.00	0.00	0.00	0.00	0.00	0.02	0.00
(iv)	F 1861#	0.51	0.00	5.04	0.00	0.00	0.00	2.42	0.00
(v)	F 1378	2.25	2.90	8.03	13.05	4.14	8.13	5.26	9.93
(vi)	F 846	0.64	3.69	0.65	0.00	0.00	0.00	0.50	0.80
(vii)	LH 1556#	0.14	0.00	2.24	0.00	0.37	13.55	1.13	2.57
Extent of adoption w.r.t % farmers##									
(i)	Ankur 651#	8	20	18	0	24	25	17	15
(ii)	White Gold#	4	0	0	0	8	0	4	0
(iii)	LHH 144#	2	0	0	0	0	0	1	0
(iv)	F 1861#	2	0	14	0	0	0	5	0
(v)	F 1378	4	0	12	0	2	15	6	0
(vi)	F 846	18	20	26	35	12	25	18	23
(viii)	LH 1556#	2	20	2	0	0	0	2	3
	n	**49**	**20**	**49**	**20**	**49**	**20**	**147**	**60**

#: Resistant varieties; ##: Multiple response.

Decimals in case of percentage of farmers rounded up to nearest whole number.

Overall in IRM villages, 34.7 and 27.7 per cent area was under *Bt* cotton (including first progeny) and hybrid cotton, respectively. The area under non-hybrid varieties was 41.5 per cent (Figure 5.18). The high percentage of are under non-hybrid varieties was due to the fact that in Ferozepur (Abohar), the majority of the farmers (60 per cent) had land holding in excess to 10 hectares. In case of non-IRM villages, the area under different cultivars is also depicted in Figure 5.18. The *desi* cotton is almost extinct as the area under it was 1.1 and 1.5 per cent in IRM and non-IRM villages, respectively.

In case of extent of adoption with respect to percentage of farmers, the similar trend was observed in all the districts. In Bathinda, Bt cotton being most widely adopted in IRM (98 per cent farmers) and non-IRM villages (80 per cent farmers) followed by hybrid varieties of American cotton (71 per cent IRM and 70 per cent non-IRM). In district Mansa, 82 and 70 per cent of the farmers in IRM and non-IRM villages had cultivated Bt cotton and 96 and 85 per cent hybrid varieties, respectively.

Table 5.14: Extent of Adoption of Different Varietal Groups During 2004

Varietal Group	*District Wise Percentage of Farmers*						*Overall Percentage of Three Districts (n=150)*	
	Bathinda (n=50)		*Ferozepur (n=50)*		*Mansa (n=50)*			
	IRM Villages	*Non-IRM Villages*	*IRM Villages*	*Non-IRM Villages*	*IRM Villages*	*Non-IRM Villages*	*IRM Villages*	*Non-IRM Villages*
Extent of adoption w.r.t. area								
American cotton								
(i) Hybrid varieties	86.7 (27.7)	28.8 (38.0)	44.0 (10.2)	10.2 (4.9)	134.6 (61.8)	30.2 (45.5)	265.3 (27.7)	69.2 (19.7)
(ii) Other varieties	30.2 (9.7)	9.4 (12.4)	346.2 (80.9)	161.8 (77.6)	21.2 (9.7)	16.6 (25.0)	397.6 (41.5)	187.8 (53.6)
(iii) *Bt*-cotton (undiscript)	134.6 (43.0)	27.2 (35.9)	30.0 (7.0)	31.2 (15.0)	43.45 (20.0)	16.0 (24.1)	208.05 (26.7)	74.4 (21.2)
(iv) *Bt*-cotton (first progeny)	53.6 (17.1)	8.8 (11.6)	5.4 (1.3)	1.6 (0.8)	18.0 (8.3)	3.6 (5.4)	77.0 (8.0)	14.0 (4.0)
Desi Cotton	7.9 (2.5)	1.6 (2.1)	2.6 (0.6)	3.6 (1.7)	0.4 (0.2)	0 (0)	10.9 (1.1)	5.2 (1.5)
Total area under cotton	313.00	75.80	428.20	208.40	217.65	66.4	958.85	350.60
Extent of adoption w.r.t. per cent farmers#								
American cotton								
(i) Hybrid varieties	71	70	35	5	96	85	67	53
(ii) Non-hybrid varieties	43	50	84	100	37	40	55	63
(iii) *Bt*-cotton (undiscript)	98	80	39	50	82	70	73	67
(iv) *Bt*-cotton (Trials)	10	0	0	0	2	0	4	0
(v) Only Bt cotton	27	5	2	0	6	5	12	3
***Desi* cotton**	12	10	4	15	2	0	6	8
n	**49**	**20**	**49**	**20**	**49**	**20**	**147**	**60**

Figures in the parenthesis are the percentage area

#: Multiple response.

Decimals rounded up to nearest whole number in case of percentage of farmers.

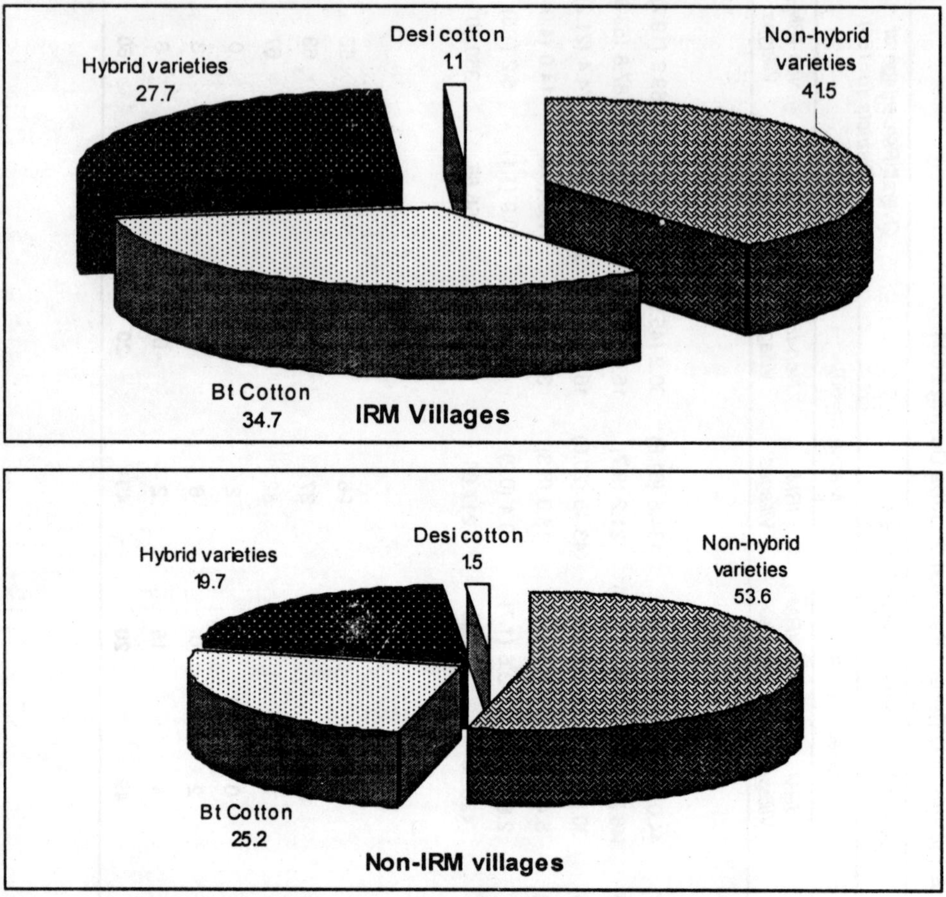

Figure 5.18: Extent of Area Under Different Cultivars of Cotton

In case of Ferozepur (Abohar) the adoption pattern was different, 84 and 100 per cent cultivated non-hybrid varieties of American cotton followed by Bt cotton 39 and 50 per cent in case of IRM and non-IRM villages, respectively (Table 5.14). In case of cotton crop, more than 60 hybrid varieties and more than 24 other varieties are cultivated in Punjab (Appendix I). The proliferation of seed companies with so many varieties has made the authenticity of different varieties doubtful. The pattern of area under different varieties also varies from district to district (Appendix I).

5.5.1.3 Most Widely Adopted Varieties

As already discussed, there was inter district diversity in the adoption of varieties. In Bathinda, 57.7 per cent of the area was under undescript Bt cotton, followed by H-8 (3.9 per cent), Vikram 11 (3.8 per cent), RCH 134 (3.4 per cent) and Ankur 651 (2.8 per cent). The detail of the varieties of cotton having area coverage of one or more than one per cent is listed in Table 5.15. In all, more than 35 varieties of cotton were cultivated in both IRM and non-IRM villages of district Bathinda. In case of Ferozepur, the distribution of area under different varieties of cotton was somewhat evenly distributed between seven varieties. The extent of adoption of F 414 was maximum (16.56 per cent), followed by Rawat Saria (12.69 per cent), Bt cotton (10.71 per cent), F 1378 (9.68 per cent), Punjab gold (8.61 per cent), Rosianwali (8.23 per cent) and SML 8 (7.82 per cent). In all, about 48 plus varieties of cotton were cultivated in IRM and non-IRM villages of Ferozepur district (Table 5.15). In district Mansa, the maximum area was under undescript Bt cotton (28.53 per cent), followed by Ankur 2226 (10.56 per cent), Ankur 651 (9.08 per cent), H 8 (7.32 per cent) and F 1378 (5.07 per cent). The area under different Ankur varieties was 25.55 per cent. In all, more than 44 varieties of cotton were cultivated in IRM and non-IRM villages of district Mansa (Table 5.15).

The scenario is likely to undergo a change with the official introduction of the Bt cotton in Punjab during 2005, the area under Bt cotton is likely to increase (Refer Table 5.49). On the surface, Bt cotton seems to offer extremely attractive means to

Table 5.15: Extent of Adoption of Different Varieties of Cotton with Respect to Area During 2004 (IRM and non-IRM villages)

District Bathinda		*District Ferozepur*		*District Mansa*	
Variety/Hybrid	*Area (%)*	*Variety/Hybrid*	*Area (%)*	*Variety/Hybrid*	*Area (%)*
Bt cotton (undiscript)	57.7	*Bt* cotton (undiscript)	10.71	*Bt* cotton (undiscript)	28.53
Ankur 651	2.8	Ankur 651	1.13	Ankur 651	9.08
RCH 134	3.4	F 1861	3.39	Ankur 2226	10.56
RCH 317	1.0	LH 1556	1.51	Ankur 2534	1.97
Dhaval 2	2.52	F 1378	9.68	Ankur 655	3.94
Birla 88	2.3	F 414	16.56	RCH 134	1.69
Navbharat 151		Rosian wali	8.23	Birla 88	1.69
Vikram 11	3.8	Rawat Saria	12.69	Nath Rani	2.39
H-8	3.9	Kiker Khera	1.57	Ganga Kaveri	1.40
Paras	1.9	PK 54	3.14	Kohinoor	2.11
Raja Sikander	2.6	F 1352	2.67	H-8	7.32
F 846	1.2	Punjab Gold	8.61	Govinda	1.69
F 1378	2.4	SML 8	7.82	Krishma	2.39
Other varieties (22+ varieties)	14.5	Other varieties (22+ varieties)	12.29	LH 1556	3.45
				F 1378	5.07
				Sikanderpuria	2.25
				Other varieties (28 + varieties)	14.47
Total area (ha)	388.8		636.6		284.05

Varieties having area coverage of one or more than one per cent are listed. Rest included in other varieties.

increase option of managing pests. In the US, around 45 per cent of cotton area was under transgenic cotton in 1998 (Norris *et al.*, 2002). Host plant resistance achieved through conventional plant breeding is probably the best IPM solution to pest problem, but use of transgenic crops, raises serious environmental and ethical questions (Norris *et al.*, 2002). Under Punjab conditions, it has to be seen whether introduction of *Bt* cotton will be an option of IPM and could be counted as an indicator of IPM adoption–only coming years will clear the air about it.

5.5.2 Adoption of Cultural Practices

Cultural methods to control insect pests involve crop husbandry practices that have a dual purpose of crop production and insect pest suppression (Litsinger 1994). A healthy, vigorous crop is one of the best forms of pest management (Norris *et al.*, 2002). There are two types of cultural control practices, primary and secondary. Primary cultural control practices are done specifically to control insect pests [*e.g.* timely (April) sowing of cotton crop to escape the damage of American bollworm], secondary cultural control practices that are specifically done for crop husbandry (*e.g.* land preparation, weeding, fertilization) but also happen to minimize insect buildup (Litsinger 1994). The cultural management of pests involves changes to the way crop is grown in order to make the crop less suitable for the pest, to make it suitable for natural enemies, or to enhance the ability of the crop to withstand pest attack (Norris *et al.*, 2002). The cultural practices disseminated under the IRM programme have been already listed under chapter 4. The different cultural management practices adopted by the IRM farmers and their comparison with non-IRM farmers are discussed here.

5.5.2.1 Level of Adoption of Recommended Seed Rate

The recommended seed rate for hybrid cotton is 1.5 kg/acre (3.75 kg ha^{-1}) and for non-hybrid varieties was 4kg/acre (10 kg ha^{-1}). Though, in case of non-hybrid varieties, the recommended seed rate given by PAU has been continuously changed from as high as 8 kg/acre (20 kg ha^{-1}) in 1979 (PAU 1979) to 3.5 kg/acre (8.75 kg ha^{-1}) in 2005 (PAU 2005). So it is difficult for farmers to keep a track of the changes in the recommendation of seed rate, and farmers also keep on experimenting with the required seed rates feasible in their area.

The results of the study indicate that almost hundred per cent of IRM and non-IRM farmers had applied lower seed rate in case of hybrid varieties of cotton, which was 60 per cent to 100 per cent lower than the recommended. The district wise detail of the level of adoption of seed rate is given in Table 5.16. In case of non-hybrid varieties majority of the farmers in IRM villages (77 per cent) and non-IRM villages (87 per cent) had applied higher seed rate. The IRM programme not having any impact in IRM villages for applying recommended seed rates is due to: one, that farmers have worked out the appropriate seed rate suitable to their area without having the consideration for the role plant density plays in pest buildup; second, the seed companies selling hybrid seed recommended 0.9kg/acre (2.250 kg ha^{-1}); third, that cost of hybrid seed is ranging between Rs. 350 to 1000 per kg and farmers are more prone to the advice of sale agents as it saves money; fourth the IRM programme had not picked up fully at the time of sowing during 2004.

Table 5.16: Level of Adoption of Seed Rate in Different Cultivars of Cotton During 2004

Adoption Level	*District Wise Percentage of Farmers*						*Overall Percentage of Three Districts (n=150)*	
	Bathinda (n=50)		*Ferozepur (n=50)*		*Mansa (n=50)*			
	IRM Villages	*Non-IRM Villages*	*IRM Villages*	*Non-IRM Villages*	*IRM Villages*	*Non-IRM Villages*	*IRM Villages*	*Non-IRM •Villages*
American cotton hybrid/varieties								
(i) Lower seed rate (750–900g/acre)	97	100	100	100	94	100	96	100
(ii) Lower seed rate (1.35g/acre)	0	0	0	0	6	0	3	0
(iii) Correct seed rate (1.5kg/acre)	3	0	0	0	0	0	1	0
n	**35**	**14**	**17**	**1**	**47**	**17**	**99**	**32**
***Bt*-Cotton (F1)#**								
(i) 450g/acre	9	0	0	50	3	7	5	15
(ii) 750–800g/acre	25	40	17	0	10	0	18	15
(iii) 900g/acre	66	60	83	50	87	93	77	70
n	**44**	**15**	**18**	**10**	**30**	**14**	**92**	**39**
Non-hybrid/varieties								
(i) Lower seed rate	19	0	0	0	6	0	6	0
(ii) Correct seed rate (3–4kg/acre)	19	20	10	10	27	13	17	13
(iii) Higher seed rate	62	80	90	90	67	87	77	87
n	**21**	**10**	**40**	**20**	**18**	**8**	**79**	**38**

#: No seed rate recommended by PAU in case of *Bt* cotton, *Bt* second progeny not included.

In case of Bt cotton, which was not recommended in Punjab state during 2004, the farmers had applied 450 g of seed per acre (by dibbling method) and 750-900 g/acre by seed cum fertilizer drill (Table 5.16).

5.5.2.2 Extent of Adoption of Timely Sowing

The manipulation of sowing time helps to minimize pest damage by producing asynchrony between host plant and the pest or synchronizing insect pests with their natural enemies or crop production with available alternate host plants of the pests or by production followed by destruction of crop residues before the insect can enter diapause (Dhaliwal and Arora 2001).

The timely sowing (April sowing) of cotton crop is one of the most important strategies of IPM, and the majority of IRM farmers were aware about this fact. The extent of timely sowing with respect to percentage of area was 51.30 per cent in IRM area and 50.39 per cent in non-IRM area (Figure 5.19). The extent of adoption of timely sowing with respect to percentage of farmers was that 26 per cent of IRM

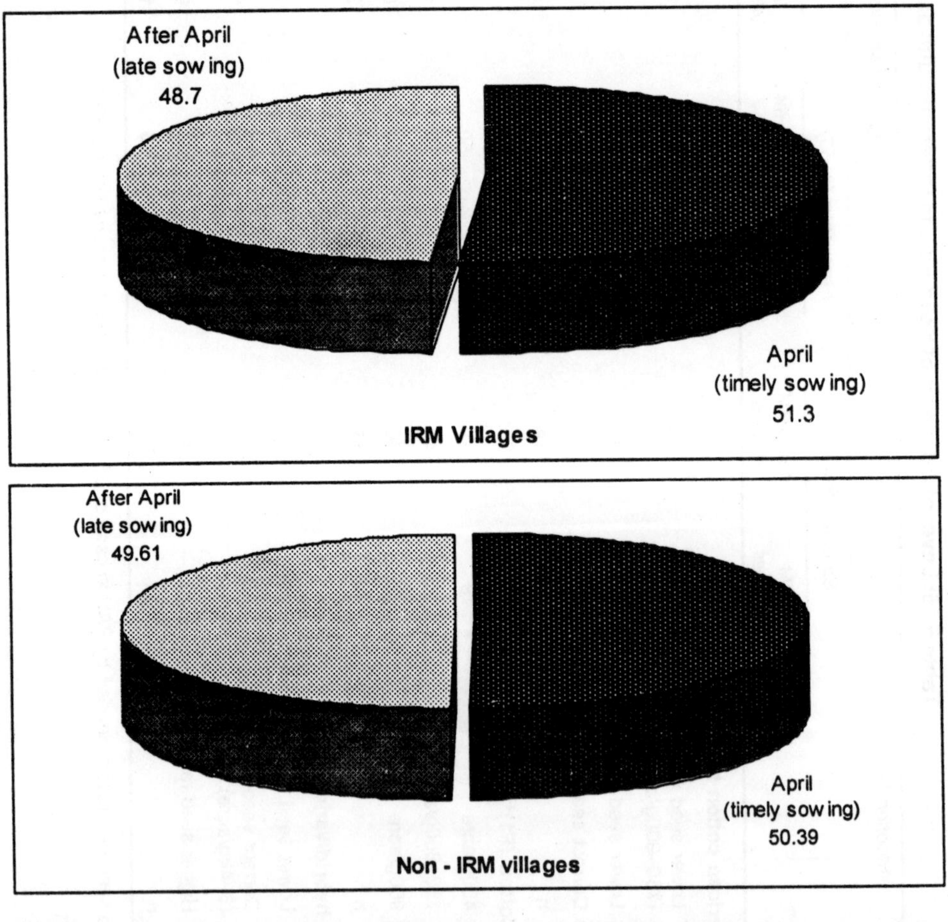

Figure 5.19: Extent of Adoption of Timely Sowing with Respect to Percentage Area

farmers had completed sowing in April, whereas 48 per cent had sown their crop April through May (Figure 5.20), which implies that 74 per cent of the IRM farmers had gone for April sowing (timely sowing) either partially or completely. In non-IRM villages 26 per cent and 52 per cent of the farmers had completely and partially, respectively, completed sowing in April. District wise detail of adoption of timely sowing with respect to percentage area and farmers is given in Table 5.17. So there is nothing to choose between IRM and non-IRM villages. It goes to show that farmers having comprehension of the benefits of timely sowing, planted their cotton crop in April and the percentage of area not covered in April was due to constraints faced by farmers and not because of any lack of knowledge and willingness on the part of farmers. The PAU, disseminating the importance of timely sowing of cotton crop through mass media and other channels has, had desired impact and the farmers having observed its multiple benefits have adopted the recommendation. The constraints in the complete adoption of timely sowing are discussed under this chapter at sub section 5.8.

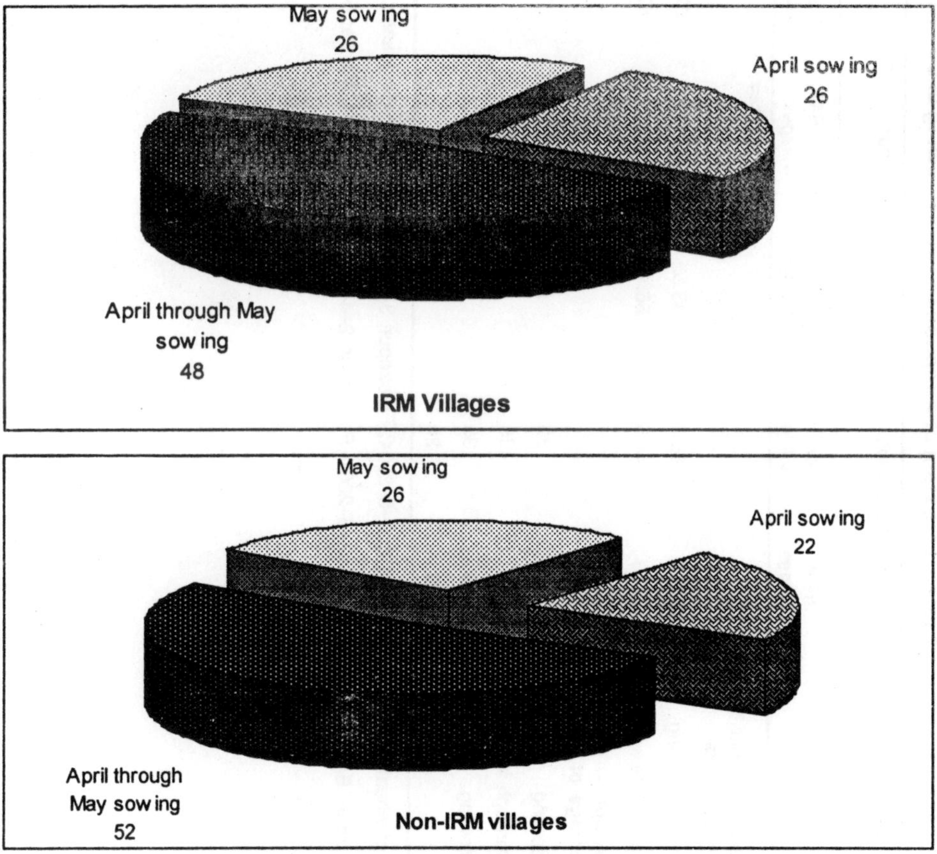

Figure 5.20: Extent of Adoption of Timely Sowing with Respect to Percentage Farmers

Table 5.17: Extent of Adoption of Timely Sowing of Cotton Crop During 2004

Month	*District Bathinda*		*District Ferozepur*		*District Mansa*		*Overall Percentage of Three Districts*	
	IRM Villages	*Non-IRM Villages*	*IRM Villages*	*Non-IRM Villages*	*IRM Villages*	*Non-IRM Villages*	*IRM Villages*	*Non-IRM Villages*
Extent of adoption w.r.t. percentage area								
1. April (timely sowing)	57.67	55.41	43.55	45.71	57.34	59.34	51.30	50.39
2. After April (late sowing)	42.33	44.59	56.45	54.29	42.66	40.66	48.70	49.61
Extent of adoption w.r.t. percentage of farmers								
(i) April sowing	22	30	20	0	35	35	26	22
(ii) April through May sowing	39	35	55	90	49	30	48	52
(iii) May sowing	39	35	25	10	16	35	26	26
n	**49**	**20**	**49**	**20**	**49**	**20**	**147**	**60**

Harvesting of wheat: Bathinda: 1–4–2004 to 2–5–2004; Ferozepur: 3–4–2004 to 10–5–2004; Mansa: 4–4–2004 to 10–5–2004.

Sowing of cotton: Bathinda: 10–4–2004 to 3–6–2004; Ferozepur: 2–4–2004 to 30–5–2004; Mansa: 6–4–2004 to 27–5–2004.

5.5.2.3 Extent of Adoption Recommended Spacing

The closer spacing in cotton results in bushy growth and affects penetration of light, hinders spraying and results in higher relative humidity that favours higher incidence of sucking pests and bollworms (Dhawan 1999). Though, high percentage of both IRM (74 per cent) and non-IRM (75 per cent) farmers had maintained proper row to row spacing in case of hybrids, but had not maintained desired plant to plant spacing. In case of line to line spacing, farmers had maintained more than the recommended spacing as they had sown with three *pora* seed drill but in case of plant to plant spacing the spacing was towards lower side of the recommendations. Similarly, in case of non-hybrid varieties, a very high percentage of farmers in IRM villages (84 per cent) and non-IRM villages (92 per cent) had maintained recommended row to row spacing (Table 5.18). There has not been any visible effect of the IRM programme on this aspect. The programme needs to go full steam even before sowing of cotton crop.

5.5.2.4 Adoption of Seed Treatment

In case of hybrid varieties, the seed purchased by farmers was already treated, but in case of non hybrid varieties, a very less percentage of five farmers had treated the seed (Table 5.18). Two farmers in IRM villages had treated their seed with cow urine plus ash of cow dung and neem (*Azadirachta indica)* leaves and reported it to be effective against termite attack. The IRM trainers must pay greater attention to seed treatment in the coming years. Some of the practices which start before sowing, and end with sowing need to be disseminated by having pre sowing group meetings. Seed treatment is one of the available tactics of IPM in Punjab against seed born diseases and to prevent early jassid attack with least disturbing the environment as well as economical.

5.5.2.5 Fertilizer Management

Excessive use of inorganic nitrogenous fertilizers generally creates congenial conditions for many insect pests. In cotton, high levels of nitrogen fertilizers result in greater attack of whitefly and bollworms (Dhaliwal and Arora 2001). Under IRM strategy, the farmers were advised to use fertilizers based on soil test and no recommendations for any dosage was given in the printed material (including IRM manual) given to the IRM farmers. The results related to extent and level of adoption fertilizers are discussed here.

In hybrid varieties, all the farmers in the IRM and non-IRM villages had applied nitrogenous and phosphatic fertilizers, and in case of non-hybrid varieties almost 89 to 100 per cent of farmers had applied these two types of fertilizers (Table 5.19). The application of muriate of potash and zinc sulphate was 16 and 22 per cent, respectively, in hybrid varieties in IRM villages and negligible in non-IRM villages. There is no hard recommendation for use of these two fertilizers in PAU, *Package Practices for Crop of Punjab*: *Kharif* as the recommendation is for applying these fertilizers on light soils (PAU 2004).

In case of Bt cotton, where there are no recommendation, a higher percentage of 21 and 18 per cent IRM farmers had applied muriate of potash and zinc sulphate

Table 5.18: Extent of Adoption of Recommended Cultural Practices in Cotton Crop During 2004

Practice	*District Wise Percentage of Farmers*						*Overall Percentage of Three Districts (n=150)*	
	District Bathinda		*District Ferozepur*		*District Mansa*			
	IRM Villages	*Non-IRM Villages*	*IRM Villages*	*Non-IRM Villages*	*IRM Villages*	*Non-IRM Villages*	*IRM Villages*	*Non-IRM Villages*
Seed Treatment	10	0	8	0	6	0	5	0
Treated seed	90	55	35	50	92	85	72	63
Seed treatment with cow urine + cow dung (ash) + neem leaves	2	0	2	0	0	0	1	0
n	**49**	**20**	**49**	**20**	**49**	**20**	**14**	**60**
Recommended spacing								
Hybrids								
(i) Line x Line	71	86	88	0	70	71	74	75
(ii) Plant x Plant	26	36	6	0	17	22	18	16
n	**35**	**14**	**17**	**1**	**47**	**17**	**99**	**32**
Non-hybrids								
(i) Line x Line	80	100	93	95	66	75	84	92
(ii) Plant x Plant	19	20	8	15	11	13	11	16
n	**21**	**10**	**40**	**20**	**18**	**8**	**79**	**38**

Decimals rounded up to nearest whole numbers.

Table 5.19: Extent of Adoption of Fertilizers and Manures in Different Cultivars of Cotton During 2004

Fertilizers/Manures Application	*District Wise Percentage of Farmers*						*Overall Percentage of Three Districts (n=150)*	
	District Bathinda		*District Ferozepur*		*District Mansa*			
	IRM Villages	*Non-IRM Villages*	*IRM Villages*	*Non-IRM Villages*	*IRM Villages*	*Non-IRM Villages*	*IRM Villages*	*Non-IRM Villages*
Hybrid varieties								
1. Nitrogenous	100	100	100	100	100	100	100	100
2. Phosphatic								
(i) DAP	94	100	100	100	100	100	100	100
(ii) Super phosphate	0	0	2	0	0	0	1	0
3. Muriate of potash	11	0	18	0	19	6	16	3
4. Zinc sulphate	14	0	18	0	30	0	22	0
n	**35**	**14**	**17**	**1**	**47**	**17**	**99**	**32**
Non-hybrid varieties								
1. Nitrogenous	100	100	100	100	89	100	97	100
2. Phosphatic								
(i) DAP	95	100	90	90	100	75	94	89
(ii) SSP	0	0	0	10	0	0	9	5
3. Muriate of potash	19	0	3	5	11	0	0	3
4. Zinc sulphate	19	0	10	20	27	0	16	11
n	**21**	**10**	**40**	**20**	**18**	**8**	**79**	**38**

Contd..

Table 5.19–Contd...

Fertilizers/Manures Application	District Wise Percentage of Farmers						Overall Percentage of Three Districts (n=150)	
	District Bathinda		District Ferozepur		District Mansa			
	IRM Villages	Non-IRM Villages	IRM Villages	Non-IRM Villages	IRM Villages	Non-IRM Villages	IRM Villages	Non-IRM Villages
***Bt* cotton**								
1. Nitrogenous	100	100	100	90	100	100	100	98
2. Phosphatic	100#	100#	95	100	98	100	99	100
3. Muriate of potash	17	13	26	0	23	14	21	10
4. Zinc sulphate	10	6	16	0	28	0	18	5
n	**48**	**16**	**19**	**10**	**40**	**14**	**107**	**40**
Urea application								
1. Basal dose	100	100	100	100	98	100	99	100
2. 1st top dressing	90	85	82	75	92	95	88	85
3. 2nd top dressing	82	85	82	70	92	95	85	85
4. 3rd top dressing	31	24	29	15	35	45	32	27
n	**49**	**20**	**49**	**20**	**49**	**20**	**147**	**60**
Farm yard manure								
1. Extent of adoption w.r.t. per cent farmers	73	90	55	40	78	40	69	57
2. Extent of adoption w.r.t. per cent farmers	22.81	9.50	10.23	6.53	19.34	13.25	16.41	8.44

#: In addition to DAP, 4 per cent and 13 per cent farmers in IRM and Non IRM villages of Bathinda, respectively had given super phosphate @ 50kg/acre

Decimals rounded up to nearest whole number; DAP: Diammonium phosphate.

compared to hybrid and non-hybrid varieties of cotton. Majority of the farmers in both the IRM and non-IRM villages had applied urea in three split doses (Table 5.19). The level of adoption of different fertilizers gives an interesting picture about the use of nitrogenous fertilizers. Against the common perception that farmers use higher nitrogenous fertilizers as in case of rice, no such thing is depicted by the data in Table 5.20. In case of hybrid varieties, recommended dosage of urea and diammonium phosphate (DAP) is 130kg/acre (325 kg ha^{-1}) and 27 kg/acre (67.5 kg ha^{-1}), respectively. Majority of the farmers, 86 per cent in the IRM villages and 94 per cent in non-IRM villages had applied lower dosages of urea (Table 5.20). In case of DAP, majority of IRM farmers (67 per cent) and non-IRM farmers (69 per cent) had applied higher dosages, mostly three times the recommended dosages. Some what similar percentage of 33 per cent (IRM farmers) and 31 per cent (non-IRM farmers) had applied correct dosage of DAP (Table 5.20).

In case of non-hybrid varieties, the recommended dosage of urea is 65kg/acre (162.5 kg ha^{-1}), which is half the dosage recommended for hybrid varieties. The farmers had used lower dosage (40 per cent), higher dosage (39 per cent) and correct dosage (21 per cent), respectively in the IRM villages (Table 5.21). In case of DAP, 51 per cent of the IRM farmers and 41 per cent of non-IRM farmers had applied correct dosages, whereas 46 per cent and 59 per cent of IRM and non-IRM farmers had applied higher dosages, respectively. The farmers did not possess the knowledge about the recommended dosage as discussed already in this chapter. Farmers had normally used half a bag (25kg), 1 bag (50kg), 1 ½ bags (75kg), 2 bags (100 kg) and so on in cotton crop. Much attention not being given to topics of fertilizer management in IPM programmes has been reported by van de Fliert (1993) and others.

The training programmes observed during the study, no meaningful training about fertilizer management, its role in IPM was discussed.

In case of *Bt* cotton, there is no PAU recommendation for fertilizer management in its *Package Practices of Kharif* 2005 even though, Bt cotton has been recommended for cultivation since 2005. The farmers had applied higher dosages of fertilizers as compared to non Bt varieties (Table 5.22). The average fertilizer dosages of urea in case of non-hybrid varieties, hybrid and Bt cotton in IRM villages was 182.7 kg, 201.80 kg and 232.90kg per hectare, respectively and somewhat similar was the pattern in non-IRM villages where in the same sequence it was 199.03 kg, 183.88 kg and 240.38 kg per hectare. Similar pattern of using higher dosage of DAP in Bt cotton is depicted in Table 5.23. In addition to higher cost of Bt cotton seed, which was almost double than the hybrid varieties and many times (almost five to six times) higher than non-hybrid varieties, the farmers had given more fertilization to Bt cotton, thus increasing the cost of cultivation of Bt cotton (excluding pesticide costs).

Farmers had also applied location specific dosages of fertilizers, as it was reported that in less fertile soil/sandy soil the fertilization given was more and in fertile soil fertilization given was less. Farmers keep on experimenting on their own for finding the optimum dosages of fertilizers suited for their area. But the importance of proper fertilization in IPM, if any, needs to be disseminated in an effective manner in IPM programmes.

Table 5.20: Level of Adoption of Fertilizer Dosages in Hybrid Varieties of Cotton During 2004

Adoption Level	District Wise Percentage of Farmers						Overall Percentage of Three Districts (n=150)	
	District Bathinda		District Ferozepur		District Mansa			
	IRM Villages	Non-IRM Villages	IRM Villages	Non-IRM Villages	IRM Villages	Non-IRM Villages	IRM Villages	Non-IRM Villages
Urea								
1. Lower dosages								
(i) 25–50 kg/acre	43	64	18	0	34	29	33	44
(ii) 50–75 kg/acre	17	7	18	0	23	24	20	16
(iii) 75–100 kg/acre	23	29	52	100	34	35	33	34
2. Correct dosages#								
(i) 125–130 kg/acre	11	0	6	0	6	12	8	6
3. Higher dosage								
(i) 150 kg/acre	6	0	6	0	2	0	4	0
n	**35**	**14**	**17**	**1**	**47**	**17**	**99**	**32**
DAP								
1. Correct dosage (25–30 kg/acre)#	39	29	18	0	34	35	33	31
2. Higher dosages								
(i) 30–40 kg/acre	9	14	0	0	13	0	9	6
(ii) 50–75 kg/acre	49	57	82	100	53	65	57	63
(iii) 100 kg/acre	3	0	0	0	0	0	1	0
n	**33**	**14**	**17**	**1**	**47**	**17**	**97**	**32**

Contd...

Table 5.20–Contd...

Adoption Level	*District Wise Percentage of Farmers*						*Overall Percentage of Three Districts (n=150)*	
	District Bathinda		*District Ferozepur*		*District Mansa*			
	IRM Villages	*Non-IRM Villages*	*IRM Villages*	*Non-IRM Villages*	*IRM Villages*	*Non-IRM Villages*	*IRM Villages*	*Non-IRM Villages*
Muriate of potash								
(i) Lower dosage 10kg/acre	0	–	0	–	11	–	6	–
(ii) Correct dosage 20–25 kg/acre	100	–	67	–	67	–	75	–
(iii) Higher dosage 50 kg/acre	0	–	33	–	22	100	19	100
n	**4**	**0**	**3**	**0**	**9**	**1**	**16**	**1**
Zinc sulphate								
(i) Correct dosage 10kg/acre	100	–	100	–	93	–	95	–
(ii) Higher dosage 20kg/acre	0	–	0	–	7	–	5	–
n	**5**	**0**	**3**	**0**	**14**	**0**	**22**	**0**

#: Urea 130kg acre^{-1}, DAP 27kg acre^{-1}–the recommended dosages.

Decimals rounded up to nearest while number. DAP: Diammonium phosphate.

Table 5.21: Level of Adoption of Fertilizer Dosages in Non-hybrid Varieties of Cotton During 2004

Adoption Level	*District Wise Percentage of Farmers*						*Overall Percentage of Three Districts (n=150)*	
	District Bathinda		*District Ferozepur*		*District Mansa*			
	IRM Villages	*Non-IRM Villages*	*IRM Villages*	*Non-IRM Villages*	*IRM Villages*	*Non-IRM Villages*	*IRM Villages*	*Non-IRM Villages*
Urea								
1. Lower dosages								
(i) 25kg/acre	10	0	0	5	0	0	2	3
(ii) 50kg/acre	52	10	23	20	56	50	38	24
2. Correct dosage#								
(i) 65–75kg/acre	19	50	20	40	25	0	21	34
3. Higher dosages								
(i) 100kg/acre	19	40	55	25	19	50	38	34
(ii) 125kg/acre	0	0	2	10	0	0	1	5
n	**21**	**10**	**40**	**20**	**16**	**8**	**77**	**28**
DAP								
1. Lower dosage (15kg/acre)	0	0	6	0	0	0	3	0
2. Correct dosage (25–30kg/acre)#	70	10	47	67	39	17	51	41
3. Higher dosage (50kg/acre)	30	90	47	33	61	83	46	59
n	**20**	**10**	**36**	**18**	**18**	**6**	**74**	**34**

Contd...

Table 5.21–Contd...

Adoption Level	*District Wise Percentage of Farmers*						*Overall Percentage of Three Districts (n=150)*	
	District Bathinda		*District Ferozepur*		*District Mansa*			
	IRM Villages	*Non-IRM Villages*	*IRM Villages*	*Non-IRM Villages*	*IRM Villages*	*Non-IRM Villages*	*IRM Villages*	*Non-IRM Villages*
Muriate of potash								
1. Correct dosage (20kg/acre)	75	–	100	–	100	–	86	–
2. Higher dosage (50kg/acre)	25	–	0	–	0	–	14	–
n	**4**	**0**	**1**	**0**	**2**	**0**	**7**	**0**
Zinc sulphate								
1. Correct dosage (10kg/acre)	100	–	100	–	80	–	92	–
2. Higher dosage (20kg/acre)	0	–	0	–	20	–	8	–
n	**4**	**0**	**4**	**0**	**5**	**0**	**13**	**0**

#: Urea 67 kg $acre^{-1}$, DAP 27 kg $acre^{-1}$

Decimals rounded up to nearest whole numbers; DAP: Diammonium phosphate.

Table 5.22: Fertilizer Dosages Applies by the Farmers in Undescript *Bt* Cotton During 2004

Fertilizer Dosages	*District Wise Percentage of Farmers*						*Overall Percentage of Three Districts (n=150)*	
	District Bathinda		*District Ferozepur*		*District Mansa*			
	IRM Villages	*Non-IRM Villages*	*IRM Villages*	*Non-IRM Villages*	*IRM Villages*	*Non-IRM Villages*	*IRM Villages*	*Non-IRM Villages*
Urea								
1. 25kg/acre	2	6	0	0	0	0	1	3
2. 50kg/acre	31	38	10	0	15	29	22	26
3. 75kg/acre	15	6	5	0	30	7	18	5
4. 100kg/acre	27	25	37	78	45	36	35	41
5. 125kg/acre	10	0	16	0	3	21	9	7
6. 150kg/acre	13	19	32	22	6	7	14	15
7. 200kg/acre	2	6	0	0	0	0	1	3
n	**48**	**16**	**19**	**9**	**40**	**14**	**107**	**39**
DAP								
1. 25–30kg/acre	40	19	11	0	41	36	35	18
2. 40kg/acre	2	12	0	0	8	0	4	5
3. 50kg/acre	52	63	50	70	49	64	50	65
4. 60kg/acre	2	0	0	0	0	0	1	0
5. 75kg/acre	0	6	11	0	2	0	3	5
6. 100kg/acre	4	0	28	30	0	0	7	7
n	**48**	**16**	**18**	**10**	**39**	**14**	**105**	**40**

Contd...

Table 5.22–Contd...

Fertilizer Dosages	*District Wise Percentage of Farmers*						*Overall Percentage of Three Districts (n=150)*	
	District Bathinda		*District Ferozepur*		*District Mansa*			
	IRM Villages	*Non-IRM Villages*	*IRM Villages*	*Non-IRM Villages*	*IRM Villages*	*Non-IRM Villages*	*IRM Villages*	*Non-IRM Villages*
Muriate of Potash								
1. 20–25kg/acre	100	100	60	–	89	–	86	50
2. 50kg/acre	0	0	40	–	11	100	14	50
n	**8**	**2**	**5**	**0**	**9**	**2**	**22**	**4**
Zinc sulphate								
1. 10kg/acre	100	100	100	–	100	–	100	100
n	**5**	**1**	**3**	**0**	**11**	**0**	**79**	**1**

No recommendation of fertilizer dosages given by PAU for *Bt* cotton.

Decimals rounded up to nearest whole numbersl; DAP: Diammonium phosphate.

Table 5.23: Average Fertilizer Dosages and FYM Applied in Different Cultivars of Cotton During 2004

Fertilizer	*District Wise Average Fertilizer Dosage (kg ha⁻¹)/FYM (t ha⁻¹)*						*Average of Three Districts (kg ha⁻¹)*	
	District Bathinda		*District Ferozepur*		*District Mansa*			
	IRM Villages	*Non-IRM Villages*	*IRM Villages*	*Non-IRM Villages*	*IRM Villages*	*Non-IRM Villages*	*IRM Villages*	*Non-IRM Villages*
Hybrid varieties								
(i) Urea	194.29	157.15	227.21	100.00	198.05	203.13	201.80	183.88
SD	84.09	67.00	66.62	0.00	66.73	66.53	73.6	69.53
n	**35**	**14**	**17**	**1**	**47**	**17**	**99**	**32**
(ii) DAP	102.27	108.33	117.65	50.00	98.61	102.95	103.29	104.90
SD	40.32	41.73	30.32	0.00	31.45	30.78	34.94	28.30
n	**33**	**14**	**17**	**1**	**47**	**17**	**97**	**32**
(iii) Muriate of potash	56.25	–	78.13	–	72.23	50.00	73.53	50.0
SD	7.23	–	31.25	–	32.95	0.00	31.23	0.00
n	**4**	**0**	**3**	**0**	**9**	**1**	**16**	**1**
(iv) Zinc sulphate	25	–	25	–	26.78	–	26.13	–
SD	0	–	0	–	6.68	–	5.33	–
n	**5**	**0**	**3**	**0**	**14**	**0**	**22**	**0**
Bt Cotton								
(i) Urea	221.68	226.58	282.9	277.78	220.83	232.15	232.90	240.38
SD	97.28	124.75	81.68	55.13	67.85	82.88	88.33	97.83
n	**48**	**16**	**19**	**9**	**40**	**14**	**107**	**39**
(ii) DAP	113.98	123.45	157.63	162.50	99.23	104.18	119.85	125.93
SD	41.98	34.13	64.75	60.38	39.00	30.50	48.75	45.8
n	48	16	18	10	40	14	106	40

Contd...

Table 5.23–Contd...

Fertilizer	District Wise Average Fertilizer Dosage (kg ha^{-1})/FYM (t ha^{-1})						Average of Three Districts (kg ha^{-1})	
	District Bathinda		District Ferozepur		District Mansa			
	IRM Villages	Non-IRM Villages	IRM Villages	Non-IRM Villages	IRM Villages	Non-IRM Villages	IRM Villages	Non-IRM Villages
(iii) Muriate of Potash	59.38	31.25	90.63	–	66.68	125	68.45	78.13
SD	5.78	–	40.03	–	22.53	–	24.25	69.38
n	**8**	**2**	**5**	**0**	**9**	**2**	**22**	**4**
(iv) Zinc Sulphate	25	25	25	–	25	–	25	25
SD	0.00	–	0.00	–	0.00	–	0.00	–
n	**5**	**1**	**3**	**0**	**11**	**0**	**19**	**1**
Non-Hybrid Varieties								
(i) Urea	154.8	206.25	208.95	200.00	164.08	187.5	182.7	199.03
SD	57.23	42.18	55.90	71.98	50.04	66.83	60.35	63.10
n	**21**	**10**	**40**	**20**	**16**	**8**	**77**	**38**
(ii) DAP	85.63	102.50	88.93	83.33	100	114.58	90.35	94.48
SD	32.35	21.90	35.2	30.33	28.65	25.53	32.98	29.38
n	**20**	**10**	**36**	**18**	**18**	**6**	**74**	**34**
(iii) Muriate of Potash	71.88	–	50.00	50	50.00	–	62.50	50.00
SD	35.9	–	0.00	0.00	0.00	–	27.95	0.00
n	**4**	**0**	**1**	**1**	**2**	**0**	**7**	**1**
(iv) Zinc Sulphate	25	–	25	28.13	30	–	26.9	28.13
SD	0	–	0	15.73	11.18	–	6.93	15.73
n	**4**	**0**	**4**	**4**	**5**	**0**	**13**	**4**
Farm yard manure	55	39	53	25	65	46	58	43
n	**36**	**18**	**27**	**8**	**38**	**8**	**101**	**34**

DAP: Diammonium phosphate.

5.5.2.6 Weed Management for Clean Cultivation

The weed management is essential to reduce build up of spotted bollworm. Carrying out sanitation tactics for clean cultivation removes over wintering inoculums and potential food source when the crop is not present (Norris *et al.*, 2002). It thus decreases the likelihood of pest carryover from season to season. The removal of weeds like *Sida* sp. (*kanghi buti*), *Abutilion* sp. (*peeli buti*) decreases the build of spotted bollworm. Under the IPM practices, it is recommended that the weeds like *Sida* sp., *Abutilon indicum*, *Datura metel* and *Xanthium* sp. (*Gutputna*) in around cotton fields should be removed on monthly basis till end June. These weeds can be removed manually or by using herbicide to reduce the build up of spotted bollworms, whitefly, and tobacco caterpillar (Dhawan *et al.*, 2004). The extent and level of adoption of different weed management practices during 2004 in IRM and non-IRM villages of Punjab are given in Table 5.24.

The method of weed management in fields, surrounding bunds and water channels is mainly physical (manual), as hundred per cent of the IRM and non-IRM farmers had adopted these methods. The extent of chemical weed control was more in IRM villages than in non-IRM villages, as 62 per cent and 40 per cent, of the farmers, respectively had adopted chemical weed control. The different herbicide used by farmers were gramoxone 24WSC, roundup 41SL and stomp 30EC. Gramoxone is the most widely used herbicide as 69 and 84 per cent of IRM and non IRM farmers respectively had applied this herbicide. Twenty two and eight per cent of IRM and non-IRM farmers, respectively, had made two applications of weedicides during the season and applied both gramoxone and roundup (Table 5.24). The average number of herbicide applications in case of district Bathinda and Mansa were higher compared to district Ferozepur. The average number of herbicide applications in IRM villages was 0.78, and 0.45 in non-IRM village (Table 5.24).

The level of adoption, of herbicide with respect to dosage in case of gramoxone was that the majority of the IRM farmers (52 per cent) had applied correct dosage (1.250 l ha^{-1}), where as in case of non-IRM farmer majority (57 per cent) had used lower dosage. In case of round up, which was applied by only 28 farmers and three farmers in IRM and non-IRM villages, respectively, 21 and 77 per cent of IRM and non-IRM farmers, respectively had applied correct dosage. Most of the IPM programmes being entomocentric (Zadoks 1989a, Peshin *et al.*, 1997), the attention towards the other aspects of IPM are peripheral. IRM programme is more entomocentric, as the name suggests, and the trainers placed a lot of emphasis on judicious use and dosages of insecticides. The trainers and the planners of training should make a concerted effort to take care of herbicide aspects in their programmes in a more effective manner than was being done during 2004. The hoeing operations, which are performed for in-row weed control, have double benefit of weed management and removing alternate host of certain insect pests or diseases. On an average, 4.58 hoeings were done by IRM farmers and 4.09 hoeings by non-IRM farmers (Table 5.24). In Ferozepur district, the average number of hoeings was 3.98 and 3.80 in IRM and non-IRM villages, respectively which is less than two other districts. The farmers cultivating non-hybrid varieties of cotton had on an average done less number

Table 5.24: Extent and Level of Adoption of Different Weed Management Practices in Cotton Crop During 2004

Practice	*District Bathinda*		*District Ferozepur*		*District Mansa*		*Total of Three Districts*	
	IRM Villages	*Non-IRM Villages*	*IRM Villages*	*Non-IRM Villages*	*IRM Villages*	*Non-IRM Villages*	*IRM Villages*	*Non-IRM Villages*
Extent of adoption of weed control practices (per cent farmers)								
1. Mechanical + manual	100	100	100	100	100	100	100	100
2. Chemical	69	50	45	20	72	50	62	40
Hoeing (per cent farmers)								
i. Manual hoeing	100	100	99	100	96	100	98	100
ii. Tratorized hoeing	100	100	100	100	100	100	100	100
iii. Average number of hoeings	4.9	4.88	3.98	3.8	4.91	3.6	4.58	4.09
SD	2.04	1.36	1.42	1.97	1.92	0.74	1.96	1.44
n	**49**	**20**	**49**	**20**	**49**	**20**	**147**	**60**
Extent of adoption of different weedicides (per cent farmers)								
1. Gramoxone 24 WSC	74	80	100	75	46	90	69	84
2. Round up 41 SL	9	0	0	0	14	10	9	4
3. Stomp 30 EC	0	0	0	25	0	0	0	4
4. Gramaxone 24 WSC + Roundup 41 SL	17	20	0	0	40	0	22	8
n	**34**	**10**	**22**	**4**	**35**	**10**	**91**	**24**

Contd...

Table 5.24–Contd...

Practice	*District Bathinda*		*District Ferozepur*		*District Mansa*		*Total of Three Districts*	
	IRM Villages	*Non-IRM Villages*	*IRM Villages*	*Non-IRM Villages*	*IRM Villages*	*Non-IRM Villages*	*IRM Villages*	*Non-IRM Villages*
Level of adoption of different weedicides (per cent farmers)								
1. Gramoxone 24 WSC								
(i) Lower dosage	22	50	32	25	33	78	29	57
(ii) Correct dosage (1.250 l ha^{-1})	52	30	45	50	57	22	52	30
(iii) Higher dosage	26	20	23	25	10	0	19	13
n	**31**	**10**	**22**	**4**	**30**	**9**	**83**	**23**
2. Roundup 41 SL								
(i) Lower dosage	56	0	–	–	89	100	79	33
(ii) Correct dosage (2.5 l ha^{-1})	44	100	–	–	11	0	21	77
n	**9**	**2**	**0**	**0**	**19**	**1**	**28**	**3**
Average number of weedicide applications	0.88	0.60	0.45	0.25	1.00	0.60	0.78	0.45
n	**49**	**20**	**49**	**20**	**49**	**20**	**147**	**60**

Percentage of farmers rounded up to the nearest whole numbers.

of hoeings, as the percentage of farmers cultivating non-hybrid varieties in district Ferozepur was 84 and 100 per cent in IRM and non-IRM villages, respectively. So the average number of hoeings was comparatively lesser than Bathinda and Mansa districts. All the farmers employ both manual and mechanical methods of hoeing. Tractorized hoeings are more common before the boll formation phase in cotton as the cost of tractorized hoeing is less than manual hoeing and requires less time.

5.5.2.7 Average Number of Other Farm Operations

The farmers on an average ploughed their cotton fields 2.7 times using cultivators and tractor drawn tillers. The ploughing was done only with tractors. The district wise average number of ploughing varied from 2.53 to 2.92. Farmers do planking after ploughing to conserve moisture and the average number of plankings were ranged between 1.79 to 2.06 and overall average was 1.94 (Table 5.25).

Table 5.25: Average Number of Ploughings, Plankings, Irrigations and Sources of Irrigation (IRM and Non-IRM villages) During 2004

	Practice	*District Bathinda*	*District Ferozepur*	*District Mansa*	*Overall*
Land preparation					
1.	Average number of ploughings	2.92	2.76	2.53	2.74
	SD	(0.94)	(0.91)	(0.73)	
2.	Average number of plankings	2.06	1.79	1.97	1.94
	SD	(0.49)	(0.45)	(0.45)	(0.43)
Irrigation					
1.	Source of irrigation (Percentage of farmers*)				
	(i) Canal	99	100	86	95
	(ii) Tube Well	100	91	97	96
	(iii) Conjugated use of canal and tube well water	38	43	22	34
2.	Average number of irrigations	7.47	5.60	5.71	6.26
	SD	(2.34)	(1.75)	(1.56)	(1.82)

Multiple response, Percentage of farmers rounded up to the nearest whole numbers.

The irrigation was given by both canal water, tube well water or conjugated use of the two. Pre-sowing irrigation was mainly with canal water as the ground water is not of good quality, being saline. The percentage of farmers using different sources of irrigation is given in Table 5.25. The average number of irrigations given was 7.74 in Bathinda, 5.71 in Mansa and 5.60 in Ferozepur. Farmers had applied more number of irrigations in Bt-cotton followed by hybrid varieties. In the villages of district Ferozepur (Abohar) adjacent to state of Rajasthan, there is shortage of irrigation water, so farmers had mostly cultivated non-hybrid varieties, and on an average given less number of irrigations.

5.5.3 Chemical Pest Control

The philosophy underlying all IPM programmes is to reduce the pesticide usage, where higher quantities of pesticides are used, and to maintain and reduce low level of pesticide applications, where the pesticide usage is less. It also varies from crop to crop. In case of cotton, very high quantities of insecticides are applied and the number of applications even touched around 30 per crop season in case of some farmers in district Ferozepur (mixtures of insecticides calculated as two or three applications depending upon the number of insecticides mixed). Change in chemical control practices by farmers trained under IPM programme(s) is an important evaluation indicator. Evaluation studies of IPM programmes have foci on change in pesticide consumption, reduction in number of pesticide applications as the main indicator of evaluation (Lacewell and Taylor 1980, NRC 1989, Pincus 1991, Winarto 1992, Head 1993, van de Fliert 1993, Peshin and Kalra 1998, Seganumullar 2000). IRM strategy is for reducing insecticide use by judicious application of insecticides, based on pest complex in cotton ecosystem rather than a single pest. The extent and level of adoption of insecticides, type of insecticides applied, average number of insecticide applications, reduction in number of tank mixed applications of insecticides, quantity of insecticides applied (technical grade material), and other related aspects are covered in this sub section.

5.5.3.1 Extent of Adoption of Different Insecticides

The extent of adoption of different insecticides is in no way, a measure for finding the impact of the IRM programmes, but it is for giving a detailed view/picture of the pattern of insecticides used in cotton crop in the study area, and can be generalised for the cotton growing belt of Punjab state. Seven groups of insecticides were used by the farmers namely, synthetic pyrethroids, organophosphates, carbamates, organochlorines, nitroguadine, oxadiazine and naturalyte. In all twenty-five insecticides belonging to these chemical groups were applied in the cotton crop (Table 5.26). In addition to these, few farmers had used fungicides and some growth regulators which have not been included in the Table 5.26.

The most widely adopted groups of insecticides were synthetic pyrethroids and organophosphates. In synthetic pyrethroids, alphamethrin was used by 56 per cent of the IRM farmers who had on an average given 3.37 applications of this insecticide. The other two most commonly used insecticides were cypermethrin (31 per cent of IRM farmers) and Fenvalerate (28 per cent of IRM farmers), who had on an average given 1.37 and 1.44 applications of these two insecticides, respectively (Table 5.26).

In the non-IRM villages the extent of adoption of synthetic pyrethroids was more as 90 per cent of farmers, had applied alphamethrin, 53 per cent had applied cypermethrin, and 23 per cent had applied fenvalerate besides other insecticides of this group. In case of organophosphates, triazophos was most commonly used insecticide, as 82 per cent and 98 per cent of the IRM and non IRM farmers, respectively had applied it followed by ethion which was applied by 63 per cent and 92 per cent

Table 5.26: Extent of Adoption of Different Insecticides During 2004

Insecticide	*District Wise Percentage of Farmers*						*Overall Percentage of Three Districts*	
	District Bathinda		*District Ferozepur*		*District Mansa*			
	IRM Villages (n=49)	*Non-IRM Villages (n=20)*	*IRM Villages (n=49)*	*Non-IRM Villages (n=20)*	*IRM Villages (n=49)*	*Non-IRM Villages (n=20)*	*IRM Villages (n=147)*	*Non-IRM Villages (n=60)*
Synthetic pyrethroids								
1. Alphamethrin 10 EC	27 (2.23)	85 (1.71)	76 (4.24)	100 (5.60)	67 (2.85)	85 (2.59)	56 (3.37)	90 (3.43)
2. Cypermethrin 10 EC	20 (1.00)	25 (1.20)	41 (1.45)	35 (1.57)	33 (1.44)	40 (5.00)	31 (1.37)	33 (2.85)
3. Cypermethrin 25 EC	4 (1.50)	5 (1.00)	10 (1.80)	0 (0.0)	12 (3.16)	55 (1.45)	9 (2.54)	20 (1.42)
4. Deltamethrin 2.8EC	6 (1.33)	0 (0.00)	8 (1.00)	20 (1.25)	8 (1.25)	10 (1.00)	7 (1.18)	10 (1.17)
5. Fenvalerate 20 EC	10 (1.00)	5 (1.00)	63 (1.55)	65 (1.54)	10 (1.2)	0 (0.00)	28 (1.44)	23 (1.50)
6. β-Cyfluthrin 0.25SC	0 (0.00)	0 (0.00)	4 (1.00)	0 (0.00)	0 (0.00)	0 (0.00)	1 (1.00)	0 (0.00)
7. Lambda cyhalothrin 5EC#	0 (0.00)	0 (0.00)	0 (0.00)	0 (0.00)	0 (0.00)	5 (2.00)	0 (0.00)	2 (2.00)
Organophosphates								
8. Profenophos 50 EC	4 (1.00)	0 (0.0)	37 (1.61)	20 (1.00)	0 (0.0)	10 (1.00)	14 (1.55)	10 (1.00)
9. Quinalphos 25 EC	24 (1.16)	0 (0.0)	24 (1.67)	10 (3.00)	16 (1.13)	15 (1.00)	22 (1.34)	8 (2.00)
10. Monocrotophos 36SL	8 (1.00)	15 (1.67)	8 (1.50)	35 (1.43)	2 (1.00)	5 (1.00)	6 (1.56)	18 (1.64)
11. Triazophos 40EC	67 (1.94)	100 (1.60)	92 (3.38)	100 (3.10)	88 (2.77)	95 (3.79)	82 (2.78)	98 (2.81)
12. Ethion 50 EC	61 (1.57)	85 (1.65)	82 (2.80)	100 (2.95)	45 (1.45)	90 (2.94)	63 (2.08)	92 (2.54)
13. Acephate 75 SP	53 (1.15)	15 (1.00)	88 (2.09)	90 (2.22)	35 (1.06)	0 (0.0)	59 (1.60)	35 (2.05)
14. Chlorpyriphos 20EC	59 (1.14)	40 (1.13)	45 (1.36)	45 (1.11)	78 (1.26)	100 (1.80)	61 (1.25)	62 (1.49)
15. Dimethoate 30EC	0 (0.00)	0 (0.0)	0 (0.0)	0 (0.00)	6 (1.00)	20 (1.00)	2 (1.00)	7 (1.00)
16. Chlorpyriphos 50EC## + Alph-amethrin 10EC	2 (1.0)	0 (0.0)	2 (1.0)	0 (0.00)	0 (0.00)	0 (0.00)	1 (1.00)	0 (0.00)

Contd...

Table 5.26–Contd...

Insecticide	*District Wise Percentage of Farmers*						*Overall Percentage of Three Districts*	
	District Bathinda		*District Ferozepur*		*District Mansa*			
	IRM Villages (n=49)	*Non-IRM Villages (n=20)*	*IRM Villages (n=49)*	*Non-IRM Villages (n=20)*	*IRM Villages (n=49)*	*Non-IRM Villages (n=20)*	*IRM Villages (n=147)*	*Non-IRM Villages (n=60)*
17. Triazophos 35 EC + Delta-methrin 1EC (Spark) ##	0 (0.00)	0 (0.0)	0 (0.0)	0 (0.00)	0 (0.00)	2 (1.00)	0 (0.00)	1 (1.00)
18. Malathion 50EC	0 (0.0)	0 (0.0)	0 (0.0)	5 (1.00)	0 (0.0)	0 (0.0)	0 (0.0)	1 (1.00)
Carbamate								
19. Thiodicarb 75WP	0 (0.00)	0 (0.00)	0 (0.00)	0 (0.00)	8 (1.00)	15 (1.00)	3	5 (1.00)
Organochlorinated								
20. Endosulfan 35EC	24 (1.17)	25 (1.20)	57 (1.39)	25 (1.40)	35 (1.00)	25 (1.00)	39 (1.25)	25 (1.20)
Nitroguadine								
21. Acetamiprid 20 SP	96 (1.73)	100 (1.55)	55 (1.15)	65 (1.15)	92 (1.78)	75 (1.33)	80 (1.62)	80 (1.35)
22. Imidacloprid 17.8SL	33 (1.13)	15 (1.40)	37 (1.50)	50 (1.20)	24 (1.17)	10 (1.00)	31 (1.28)	25 (1.40)
23. Thiomethoxam 25 WSC	47 (1.39)	70 (1.00)	2 (1.00)	5 (1.00)	0 (0)	20 (1.00)	16 (1.38)	32 (1.05)
Oxadiazine								
24. Indoxacarb 15SC	47 (1.48)	50 (1.50)	37 (1.22)	65 (1.08)	67 (1.48)	40 (1.25)	50 (1.42)	52 (1.26)
Naturalyte								
25. Spinosad 48 SC	31 (1.07)	5 (2.00)	37 (1.33)	70 (1.43)	39 (1.26)	35 (1.43)	35 (1.23)	37 (1.46)

Figures in the parenthesis are the average frequency of a particular insecticide applied by the adopters.

Decimals in Percentage of farmers rounded up to nearest whole number.

#: Not recommended; ##: Ready made cocktails; Multiple response.

of the IRM and non-IRM farmers, respectively. The average number of applications of triazophos given by adopters was 2.78 in IRM villages and 2.81 in non-IRM villages, and in case of ethion 2.08 and 2.54 in IRM and non-IRM villages, respectively. The other most commonly used insecticides in organophosphates were chlorpyriphos (61 per cent IRM farmers and 62 per cent non IRM farmers), acephate (59 per cent IRM farmers and 35 per cent non-IRM farmers). The farmers in non-IRM villages had on an average given more applications of different organophosphate insecticides than IRM farmers (Table 5.26). One per cent of farmers in both IRM and non-IRM villages had applied readymade cocktails of chlorpyriphos + alphamethrin, and triazophos + deltamethrin, respectively, which are readily available in the market inspite of the fact that these are not being recommended in Punjab. This shows the regulatory mechanism of pesticides is faulty.The nitroguadine group is also widely used in cotton crop. In this group acetamiprid was widely used by 80 per cent of farmers, in IRM and non-IRM villages who on an average gave 1.62 and 1.35 applications, respectively of this insecticide. The other insecticides of this group imidacloprid and thiomethoxam were used by 31 and 16 per cent of IRM farmers, respectively and 25 and 32 per cent of non-IRM farmers, respectively. The insecticides of this group were used against sucking pests namely jassid and whitefly. The carbamates were applied by three and five per cent of IRM and non IRM farmers, respectively.

The use of endosulfan increased in IRM villages, as it was the only chemical recommended for use during 60-90 days after sowing of cotton crop, instead of monocrotophos. The endosulfan is considered relatively safe to natural enemies, whereas monocrotophos is highly toxic to natural enemies under IRM strategy. More percentage of IRM farmers had applied it as the first spray before 90 days after sowing. In non-IRM villages, it was applied during later stages (after 90 days after sowing), and in IRM villages, it was also applied for control of tobacco caterpillar after 90 days after sowing. Even the insecticide company *'Excel'* producing this molecule, brand it as IPM compatible insecticide, thus propagating their product as IPM product.

The other two new groups of insecticides *viz.* oxadiazine (indoxacarb) and naturalyte (spinosad) have been introduced in Punjab in 2003, which are costlier and the cost of insecticide per hectare, in case of former is about Rs. 1400 (@ 0.5 l ha^{-1}), and in case of latter about Rs. 1500 (@ 150 ml ha^{-1}). These two insecticides are mainly recommended for *Helicoverpa armigera* (ABW) and the farmers having observed the good results in controlling ABW in 2003, had applied these insecticides. These insecticides being new molecules, so there is no resistance in ABW against these insecticides. The percentage of IRM farmers, having applied indoxacarb was 50 per cent, and non IRM farmers 52 per cent with an average number of applications 1.42 and 1.26, respectively. Spinosad was applied by 35 per cent IRM farmers and 37 per cent non-IRM farmers with an average number of applications of 1.23 and 1.46, respectively. District wise detail of different insecticides applied by the percentage of farmers is given in Table 5.26.

5.5.3.2 Level of Adoption of Different Insecticides

Dosage plays an important role in IPM programmes. Proper dosage and dilution is important in IRM to prevent development of pesticide resistance in insect pests and pest resurgence. In many IPM farmer field school programmes it has been observed, that no attention was paid to correct pesticide application (van de Fliert 1993). But under the IRM training programme enough attention was paid to use of insecticides according to good agricultural practices *i.e.* right timing–use insecticides only when needed, right chemical–choose appropriate insecticide, right dosage–use only recommended dosage, and right method–use proper sprayer and spray method. This was done week after week, through group discussions, street plays, leaflets, information centres and scouts.

The level of adoption, with respect to dosages as recorded by the farmers in the questionnaire file given to them at the start of cropping season in 2004, revealed interesting information. This method of giving farmers the questionnaire at the start of cropping season was to avoid any difficulty in terms of recall or recollect of dosages, concentrations and dates of application, through retrospective-interview. The farmers had not applied same dosages of a particular insecticide throughout the cropping season in case of alphamethrin, cypermethrin, fenvalerate, acephate, chlorpyriphos, ethion, monocrotophos, profenophos, triazophos and imidacloprid. The farmers in both IRM and non-IRM villages had used lower dosages of insecticides, normally when mixing the two insecticides and higher dosages when spraying a particular insecticide without mixing. In the IRM villages, the percentage of farmers using correct dosages of different insecticide was more than non-IRM villages. Percentage of farmers in IRM and non-IRM villages, respectively using correct dosages in case of alphamethrin was 29 and 9 per cent, cypermethrin eight and zero per cent, fenvalerate 10 and zero per cent, acephate 76 and 57 per cent, chlorpyriphos 51 and 46 per cent, ethion 67 and 64 per cent, monocrotophos 78 and 46 per cent, profenophos 75 and 50 per cent, quinalphos 75 and 100 per cent, triazophos 64 and 53 per cent, acetamiprid 11 and seven per cent, indoxacarb 95 and 94 per cent and spinosad 33 and nine per cent. However, in case of five insecticides more percentage of non-IRM farmers had applied correct dosages than IRM farmers. These insecticides were cypermethrin, deltamethrin, endosulfan, thiomethoxam and imidacloprid. The farmers in IRM villages were also using lower or higher dosage of insecticides. More effort on the part of trainers to make farmers understand the importance of dosages in IPM to avoid build up of resistance in insect pests. The farmers must be asked to refer the manual given to them for dosages, as it is difficult to recollect the knowledge of specifics, when there is large number of insecticides recommended to choose from. The detail regarding lower, correct and higher dosages of different insecticides applied is given in Table 5.27. The farmers had used higher concentrations in case of grown-up larvae and lower concentration in case of young larvae of American bollworm. The concentrations also varied according to crop stage. In the IRM training, attention must be paid to concentration of insecticide solutions.

5.5.3.3 Extent of Adoption of ETL

Crop monitoring is an important aspect of IPM for assessing the pest build up, to observe the natural enemies and other aspects of crop cultivation. The farmers made regular visit to their fields. Mostly the farmers visited a particular field twice a week and small farmers on daily basis. The other objective of regular field visits is to find out economic threshold level of different insect pests, and based on that make the

Table 5.27: Level of Adoption of Different Insecticides During 2004

Insecticide	District Wise Percentage of Farmers						Overall Percentage of Three Districts	
	District Bathinda		District Ferozepur		District Mansa			
	IRM Villages	Non-IRM Villages	IRM Villages	Non-IRM Villages	IRM Villages	Non-IRM Villages	IRM Villages	Non-IRM Villages
Alpha-methrin 10EC								
(i) Correct dosage (250 ml/ha)	23[#]	0	14[#]	0	48[#]	29[#]	29[#]	9[#]
(ii) Higher dosage	85[#]	100	100[#]	100	58[#]	100[#]	81[#]	100[#]
n	**13**	**17**	**37**	**20**	**33**	**17**	**83**	**54**
Cypermethrin 10EC								
(i) Lower dosage	10[#]	0	5	0	31[#]	0	15[#]	0
(ii) Correct dosage (500 ml/ha)	100[#]	100	90	100	56[#]	100	80[#]	100
(iii) Higher dosage	0	0	5	0	19[#]	0	9[#]	0
n	**10**	**5**	**20**	**7**	**16**	**8**	**46**	**20**
Cypermethrin 25EC								
(i) Lower dosage	0	0	0	–	33	0	15	0
(ii) Correct dosage (200 ml/ha)	50	0	0	–	0	0	8	0
(iii) Higher dosage	50	100	100	–	67	100	77	100
n	**2**	**1**	**5**	**0**	**6**	**11**	**13**	**12**
Delta-methrin 2.8EC								
(i) Correct dosage (400 ml/ha)	33	–	25	75	100	100	55	83
(ii) Higher dosage	67	–	75	25	0	0	45	17
n	**3**	**0**	**4**	**4**	**4**	**2**	**11**	**6**

Contd...

Table 5.27–Contd...

Insecticide	District Wise Percentage of Farmers						Overall Percentage of Three Districts	
	District Bathinda		District Ferozepur		District Mansa			
	IRM Villages	Non-IRM Villages	IRM Villages	Non-IRM Villages	IRM Villages	Non-IRM Villages	IRM Villages	Non-IRM Villages
Fenvalerate 20EC								
(i) Correct dosage (250ml ha^{-1})	0	0	13#	0	0	–	10#	0
(ii) Higher dosage	100	100	90#	100	100	–	93#	100
n	**5**	**1**	**31**	**13**	**5**	**0**	**41**	**14**
β-cyfluthrin 0.25EC								
(i) Correct dosage (500 ml ha^{-1})	–	–	0	–	–	–	0	–
(ii) Higher dosage	–	–	100	–	–	–	100	–
n	**0**	**0**	**2**	**0**	**0**	**0**	**2**	**0**
Lambda cyhalothrin 5EC##								
(i) 1.200ml ha^{-1}	100	–	–	–	100	100	100	100
n	**2**	**0**	**0**	**0**	**1**	**1**	**3**	**1**
Acephate 75SP								
(i) Lower dosage	8	0	44#	61#	35	–	31#	52#
(ii) Correct dosage (2 l ha^{-1})	92	100	70#	50#	65	–	76#	57#
(iii) Higher dosage	0	0	7#	6#	0	–	3#	5#
n	**26**	**3**	**43**	**18**	**17**	**0**	**86**	**21**
Chlorpyriphos 20EC								
(i) Lower dosage	79#	100	68#	56	24	35	53#	54
(ii) Correct dosage (5l ha^{-1})	24#	0	41#	44	76	65	51#	46
n	**29**	**8**	**22**	**9**	**38**	**20**	**89**	**37**

Table 5.27–Contd...

Insecticide	*District Wise Percentage of Farmers*						*Overall Percentage of Three Districts*	
	District Bathinda		*District Ferozepur*		*District Mansa*			
	IRM Villages	*Non-IRM Villages*	*IRM Villages*	*Non-IRM Villages*	*IRM Villages*	*Non-IRM Villages*	*IRM Villages*	*Non-IRM Villages*
Dimethoate 30EC								
(i) Correct dosage (625 ml/ha^{-1})	–	–	–	–	–	–	0	0
(ii) Higher dosage	–	–	–	–	100	100	100	100
n	**0**	**0**	**0**	**0**	**3**	**4**	**3**	**4**
Ethion 50EC								
(i) Lower dosage	13#	0	65#	70#	32	39#	40#	38#
(ii) Correct dosage (2l ha^{-1})	93#	100	50#	35#	64	61#	67#	64#
(iii) Higher dosage	3#	0	8#	0	4	6#	5#	2#
n	**30**	**17**	**40**	**20**	**22**	**18**	**92**	**55**
Monocrotophos 36SL###								
(i) Lower dosage	0	0	50#	43	0	0	22#	27
(ii) Correct dosage (1.5 l ha^{-1})	75	33	75#	57	100	0	78#	46
(iii) Higher dosage	25	67	0	0	0	100	11#	27
n	**4**	**3**	**4**	**7**	**1**	**1**	**9**	**11**
Profenophos 50EC								
(i) Lower dosage	0	–	6	0	–	0	5	0
(ii) Correct dosage (1.25 ha^{-1})	100	–	72	25#	–	100	75	50#
(iii) Higher dosage	0	–	22	100#	–	0	20	67#
n	**2**	**0**	**18**	**4**	**0**	**2**	**20**	**6**

Contd...

Table 5.27–Contd...

Insecticide	*District Wise Percentage of Farmers*						*Overall Percentage of Three Districts*	
	District Bathinda		*District Ferozepur*		*District Mansa*			
	IRM Villages	*Non-IRM Villages*	*IRM Villages*	*Non-IRM Villages*	*IRM Villages*	*Non-IRM Villages*	*IRM Villages*	*Non-IRM Villages*
Quinalphos 25EC								
(i) Lower dosage	17	–	0	0	38	0	16	0
(ii) Correct dosage (2l ha^{-1})	66	–	92	100	62	100	75	100
(iii) Higher dosage	17	–	8	0	0	0	9	0
n	**12**	**0**	**12**	**2**	**8**	**3**	**32**	**5**
Triazophos 40EC								
(i) Lower dosage	39#	0	47#	90#	35#	63#	40#	51#
(ii) Correct dosage (1.5l ha^{-1})	64#	70#	62#	15#	65#	74#	64#	53#
(iii) Higher dosage	21#	35#	13#	5#	21#	21#	18#	20#
n	**33**	**20**	**45**	**20**	**43**	**19**	**121**	**59**
Thiodicarb 75WP								
(i) Correct dosage (625ml ha^{-1})	–	–	–	–	0	–	0	–
(ii) Higher dosage	–	–	–	–	100	–	100	–
n	**0**	**0**	**0**	**0**	**4**	**0**	**4**	**0**
Endosulfan 35EC								
(i) Lower dosage	42	40	32	20	35	0	35	20
(ii) Correct dosage (2.5 l ha^{-1})	58	60	68	80	41	40	58	60
(iii) Higher dosage	0	0	0	0	24	60	7	20
n	**12**	**5**	**28**	**5**	**17**	**5**	**57**	**15**

Contd...

Table 5.27–Contd...

Insecticide	*District Wise Percentage of Farmers*						*Overall Percentage of Three Districts*	
	District Bathinda		*District Ferozepur*		*District Mansa*			
	IRM Villages	*Non-IRM Villages*	*IRM Villages*	*Non-IRM Villages*	*IRM Villages*	*Non-IRM Villages*	*IRM Villages*	*Non-IRM Villages*
Imidacloprid 17.8SL								
(i) Correct dosage (100 ml ha^{-1})	76#	60	41	54	51#	73	58#	63
(ii) Higher dosage	36#	40	59	46	58#	27	50#	37
n	**45**	**20**	**27**	**13**	**45**	**15**	**117**	**48**
Acetamiprid 20SP								
(i) Correct dosage (50 gm ha^{-1})	25	33	6	0	0	0	11	7
(ii) Higher dosage	75	67	94	100	100	100	89	93
n	**16**	**3**	**18**	**10**	**12**	**2**	**46**	**15**
Thiomethoxam 25WSC								
(i) Lower dosage	4	0	0	0	–	0	4	0
(ii) Correct dosage (100 gm ha^{-1})	44	36	100	100	–	100	46	53
(iii) Higher dosage	52	64	0	0	–	0	50	47
n	**23**	**14**	**1**	**1**	**0**	**4**	**24**	**19**
Indoxacarb 15SC								
(i) Lower dosage	4	20	11	0	0	0	4	6
(ii) Correct dosage (500 ml ha^{-1})	96	80	83	100	100	100	95	94
(iii) Higher dosage	0	0	6	0	0	0	1	0
n	**23**	**10**	**18**	**13**	**33**	**8**	**74**	**31**

Contd...

Table 5.27–Contd...

Insecticide	*District Wise Percentage of Farmers*						*Overall Percentage of Three Districts*	
	District Bathinda		*District Ferozepur*		*District Mansa*			
	IRM Villages	*Non-IRM Villages*	*IRM Villages*	*Non-IRM Villages*	*IRM Villages*	*Non-IRM Villages*	*IRM Villages*	*Non-IRM Villages*
Spinosad 48SC								
(i) Lower dosage	13	0	6#	0	5	0	8#	0
(ii) Correct dosage (150 ml ha^{-1})	40	0	28#	14	32	0	33#	9
(iii) Higher dosage	47	100	78#	86	63	100	63#	91
n	**15**	**1**	**18**	**14**	**19**		**52**	**22**

Decimals have been rounded up to nearest whole number.

#: Farmer applied different dosage of a particular insecticide for spraying on different occasions; ##: Not recommended by PAU
###: Not recommended under IRM strategy.

pesticide use decision. In the USA and other developed countries, where farmers have large landholdings compared to this part of the world, farmers hire the services of scouts for monitoring their crop and finding threshold levels. There are different schools of thought on the applicability of ETL. One school of thought is that there are technical flaws in the ETL concept, namely:

1. Prices for harvested product and pesticides tend to fluctuate over the season.
2. Natural enemy populations are not considered in the ETLs.
3. Many other factors that farmers usually consider in pest control decision making are not included in the ETL concept. For instance, weather conditions, vigour of crop, availability of irrigation water, availability of cash and credit facilities, other investment opportunities and indigenous practices (Zadoks 1985, Bently and Andrews 1996, van de Fliert 1998).

The second school of thought is the adoptability of ETL methodology at farm level as:

1. The observation techniques required to obtain a representative sample are often complex and laborious.
2. Pest population assessment methods apply units that farmers may not be familiar with, such as percentages, square meter, hectare.
3. An ETL applies for one pest, whereas farmers face a complex of pests in their crop having to consider more than one ETL at a time is too complicated (van de Fliert 1993, Bently and Andrews 1996).

The third school of thought, the entomologists consider ETL as the basic requirement for making pesticide decisions. Thresholds have been considered the first step to transition to level one IPM along with pest detection, and had been incorporated in all IPM programmes (Bhuiyan *et* al., 1992, Kogan and Bajwa 1999).

Under the IRM programme in Punjab, the ETL of insect pests have been recommended for jassid, whitefly, and bollworm complex. ETL for jassid was recommended by PAU in 1979 (PAU 1979), whitefly and bollworm in 1991 (PAU 1991), but none of the farmers in non-IRM villages had used ETL for deciding the pesticide application. In IRM villages, where the trainers and scouts were practically showing the method of calculating ETL, only 44 per cent of the farmers had applied ETL concept. Even though the finding of ETL in case of jassid is much simpler as the farmer has to observe whether on 50 per cent of plants curling along leaf margins has taken place, yet the adoption was only 33 per cent during the IRM programme. In case of whitefly, the farmers' have to survey crop twice a week and check three upper leaves of plant, if there are 6 adults per leaf on surveyed plant, the insecticide has to be applied. For American bollworm and spotted bollworm, field has to be roughly divided into four parts and from these four parts freshly shed fruiting bodies are to be collected, if 5 per cent or more are damaged, insecticides have to be applied. The IRM farmers adopting ETL was 33 per cent in case of American bollworm, 17 per cent in case of spotted bollworm and 25 per cent in case of whitefly and the overall percentage of adoption of ETL for different insect pests was 44 per cent (Table 5.28). More than

50 per cent of the farmers, in IRM villages did not adopt ETL for any of the insect pests for making pesticide use decision.

Table 5.28: Extent of Adoption of Economic Threshold Level (ETL)

Category	*District Wise Percentage of IRM Farmers*			*Overall Percentage of Three Districts*
	Bathinda	*Ferozepur*	*Mansa*	*IRM*
Adoption of ETL w.r.t percentage of farmers	55	37	39	44
Adoption of ETL for a particular pest				
(i) American bollworm	41	27	31	33
(ii) Spotted bollworm	23	12	16	17
(iii) Jassid	45	24	31	33
(iv) Whitefly	29	22	24	25
Number of pests for which ETL adopted				
(i) For 4 pests	16	0	6	7
(ii) For 3 pests	12	14	18	15
(iii) For 2 pests	18	14	9	14
(iv) For 1 pest	9	9	6	8
(v) No pest	45	63	61	56
n	49	49	49	147

In non-IRM villages none of the farmers adopted ETL.
Decimals rounded up to the nearest whole number.

The study confirms the fact that diffusion of complex technologies through package of practices or other modes of extension as used for the green revolution technologies like high yielding varieties, seeds etc. does not work. Since the introduction of ETL concept of Jassid in 1979 by PAU, 25 years have gone by, but rate of adoption is zero and same is the case with the ETL of whitefly and bollworms. The extension sub-system of the PAU has failed to provide any feedback to the research sub-system. Whether the ETL concept will be adopted in the IRM villages once the intervention is withdrawn, after 2007, needs to be studied. The adoptability of ETL and the constraints have been discussed in this chapter at sub section 5.7.5.

5.5.3.4 Cocktails of Insecticides: Perception and Practice

Along with rotation of different molecules of different insecticides in a season and not to use insecticides of same group in consecutive sprays, another important strategy to avoid build up resistance in insect pests is, not to mix the insecticides. The PAU, Ludhiana has been propagating/disseminating this strategy of no tank mixing since long but the ground level impact of it has been almost zero. The farmers before the IRM programme in pre-training knowledge test reported that tank mixing is desirable as 93 per cent reported that insecticides should be tank mixed and only seven per cent reported that insecticides should not be tank mixed. The farmers perceptions about tank mixing is that it gives better results in controlling the insect pests than spraying single insecticides especially in case of bollworms or when two pests are present at the same time, and it is advisable to spray tank mixtures than to spray separately for two pests. In IRM villages, farmers had the knowledge that insecticides should not be tank mixed as the same was reported by 91 per cent IRM farmers post-training. Only nine per cent, of the IRM farmers' perception of no tank mixing had not changed. But whether the change in perception in terms of gain in knowledge (lowest level of cognitive domain) translated into practice is given in Table 5.29.

Farmers in IRM villages had applied 33 different combinations of tank mixtures and tank mixing pattern varied from district to district and village to village, indicating that local factors, like the influence of other farmers etc. affect these decisions. In district Bathinda, 12 different combinations of insecticides were mixed in five IRM villages. In five IRM villages of Ferozepur, 25 different combinations were used, whereas in five IRM villages of Mansa 13 different cocktails were used. The most common cocktails applied were alph-amethrin with triazophos (181 in IRM and 97 in non-IRM villages) followed by alpha-methrin with ethion (93 in IRM and 87 in non-IRM villages). The number being higher in IRM villages is due to larger sample size 147 against 60 in non-IRM villages. The detail of different cocktails of insecticides used is given in Appendix V. In district Bathinda, very less number of cocktails were applied compared to other districts and the reasons for that are that IRM-farmers in Bathinda had higher knowledge about different IPM practices, the area under Bt cotton was more and the average number of insecticide applications were less than in other districts.

Reduction in the use of tank mixtures is another reward of IRM programme as the IRM farmers in all the districts had comparatively used less number of tank mixtures than non-IRM farmers. IRM farmers had applied 30.81 per cent of the total insecticide application as tank mixtures, whereas non IRM farmers had applied one out of every two insecticide applications as tank mixtures Non-IRM farmers had applied 50.16 per cent of the total insecticide applications as tank mixtures (Table 5.29). The percentage of IRM farmers using tank mixtures was 64, where as 95 per cent non-IRM farmers had applied tank mixtures.

5.5.3.5 Average Number of Insecticide Applications

The reduction in the frequency of pesticide applications is an important indicator of the benefits of IPM programme as discussed in the beginning of this section.

The farmers had recorded the use of different insecticides against a particular insect pest in the questionnaire and on that basis the average number of insecticide applications against a particular insect pest was calculated. The maximum number of insecticide applications was used against bollworm complex and especially against American bollworm even though the outbreak of this insect pest was less during 2004 compared to previous years. American bollworm became a key pest of cotton during the period 1991-95. The pest scenario on cotton in north zone during the last

Table 5.29: Frequency of Different Cocktails (Mixtures) of Insecticides Used in Cotton Crop (Not recommended under IRM strategy or by the PAU) During 2004

Insecticide Mixture	*District Wise Frequency of Tank Mixtures*						*Total Frequency of Three Districts*	
	District Bathinda		*District Ferozepur*		*District Mansa*			
	IRM Villages (n=49)	*Non-IRM Villages (n=20)*	*IRM Villages (n=49)*	*Non-IRM Villages (n=20)*	*IRM Villages (n=49)*	*Non-IRM Villages (n=20)*	*IRM Villages (n=147)*	*Non-IRM Villages (n=60)*
Alpha-methrin 10 EC with:								
(i) Acephate 75 SP	0	0	7	12	0	0	7	12
(ii) Chlorpyriphos 20EC	2	0	0	0	2	0	4	0
(iii) Endosulfan 35EC	0	0	2	0	0	0	2	0
(iv) Ethion 50EC	11	18	57	51	28	23	96	90
(v) Monocrotophos 36SL	0	0	1	0	1	0	2	0
(vi) Profenophos 50EC	0	0	10	3	0	0	10	3
(vii) Quinalphos 25EC	0	0	4	0	0	0	4	0
(viii) Triazophos 40EC	13	11	88	48	80	38	181	97
Cypermethrin with:								
(i) Acephate 75SP	0	0	1	0	2	0	3	0
(ii) Chlorpyriphos 20EC	1	0	2	0	3	4	6	4
(iii) Endosulfan 35EC	0	0	1	0	0	0	1	0
(iv) Ethion 50EC	12	5	12	0	22	34	46	46
(v) Profenophos 50EC	0	0	4	0	0	0	4	0
(vi) Quinalphos 25EC	0	0	0	3	1	0	1	3
(vii) Triazophos 40EC	1	0	9	3	17	6	27	9
(viii) Monocrotophos 36SL	0	1	1	0	0	0	1	1

Contd...

Table 5.29–Contd...

Insecticide Mixture	*District Wise Frequency of Tank Mixtures*						*Total Frequency of Three Districts*	
	District Bathinda		*District Ferozepur*		*District Mansa*			
	IRM Villages (n=49)	*Non-IRM Villages (n=20)*	*IRM Villages (n=49)*	*Non-IRM Villages (n=20)*	*IRM Villages (n=49)*	*Non-IRM Villages (n=20)*	*IRM Villages (n=147)*	*Non-IRM Villages (n=60)*
Delta-methrin 2.8EC with:								
(i) Endosulfan 35EC	0	0	1	0	0	0	1	0
(ii) Ethion 50EC	0	0	1	0	0	0	1	0
(iii) Triazophos 40EC	0	0	0	8	2	3	2	11
Fenvalerate 20EC with:								
(i) Acephate 75SP	0	0	21	10	1	0	22	10
(ii) Endosulfan 35EC	0	0	1	0	0	0	1	0
(iii) Ethion 50EC	1	0	6	4	0	0	7	4
(iv) Monocrotophos 36SCL	0	0	2	6	0	0	2	6
(v) Profenophos 25EC	0	0	1	0	0	0	1	0
(vi) Quinalphos 25EC	0	0	3	4	0	0	3	4
(vii) Triazophos 40EC	1	0	2	0	0	0	3	0
Lambda cyhalothrin +								
Ethion 50EC	2	0	0	0	0	0	2	0

Contd...

Table 5.29–Contd...

Insecticide Mixture	*District Wise Frequency of Tank Mixtures*						*Total Frequency of Three Districts*	
	District Bathinda		*District Ferozepur*		*District Mansa*			
	IRM Villages (n=49)	*Non-IRM Villages (n=20)*	*IRM Villages (n=49)*	*Non-IRM Villages (n=20)*	*IRM Villages (n=49)*	*Non-IRM Villages (n=20)*	*IRM Villages (n=147)*	*Non-IRM Villages (n=60)*
Acephate 75SP +								
Profenophos 50EC	2	0	0	0	2	0	4	0
Ethion 50EC +								
Indoxacarb 15SC	1	5	0	0	0	0	6	0
Triazophos 40EC with:								
(i) Chlorpyriphos 36SL	0	2	0	0	1	0	3	0
(ii) Ethion 50EC	0	0	0	0	0	9	0	9
Monocrotophos + Ethion + Alpha-methrin	0	0	0	0	0	4	4	0
Ethion + Triazophos + Alpha-methrin	2	0	0	0	0	0	2	0
Endosulfan + Ethion + Alph-amethrin	0	0	0	1	0	0	0	1
Triazophos + Ethion + Fenvalerate	0	0	0	1	0	0	0	1
Total frequency of tank mixtures of insecticides	49	35	244	154	162	121	455	311
Percentage insecticide applications as tank mixtures#	12.28	25.18	38.07	58.78	34.18	56.22	30.81	50.24

#: Based on total number of insecticide applications by all the respondent farmers.

three decades is given in Appendix VI. In 2004, the tobacco caterpillar which was a minor pest up to 2003 emerged as the major pest during 2004. Spotted bollworm infestation was more during 2004 in comparison to previous two years.

The average number of insecticide applications against American bollworm and spotted bollworm were 8.50 in IRM villages compared to 11.5 in non-IRM villages. This was due to the fact that IRM farmers had made informed decisions in applications of insecticides against bollworms. The average number of insecticide applications against tobacco caterpillar, whitefly, jassid were 1.55, 1.46, 1.47 respectively in IRM villages, whereas in non-IRM villages against average number of insecticide applications against tobacco caterpillar, whitefly and jassid were 1.23, 1.21 and 1.80, respectively (Table 5.30). Farmers had applied acephate also for growth of the plant in the areas where there was shortage of irrigation water, as the farmers perception was that it enhances the growth of the plant, which is not a better practice. The trainers were seen explaining to the farmers that insecticides should not be applied for growth purpose, but still on an average 0.05 number of applications were given by farmers for the same.

The average number of insecticide applications was compared with the base line data of 2003. During baseline data farmers were not able to give an exact number of tank mixtures of insecticides applied during 2003. Therefore, for comparing the average number of insecticide applications of 2003 and 2004, every mixture of insecticides used during 2004 was calculated for as one application. There was a significant reduction in average number of insecticide applications during 2004 in comparison to 2003 at $p<0.01$. The reduction was in case of both IRM and non-IRM villages. The average number of insecticides in IRM and non-IRM villages was computed in two ways. One procedure adopted was, that tank mixture of two or more than two insecticides were taken as separate applications and second procedure was, that even tank mixture was taken as a single application to test our hypotheses that IRM programme resulted in lesser number of insecticide applications, lesser number of tank mixtures in IRM villages as compared to non-IRM villages.

The results indicate (Table 5.31) that average number of insecticide applications (calculating tank mixtures separately) in IRM villages were lesser than in non-IRM villages in all the three districts. In district Bathinda, the average number of insecticide applications during 2004 were 8.45 in IRM villages and 8.85 in non-IRM villages, a positive difference of 4.7 per cent but not statistically significant as calculated 't' value is 0.51 which is less than table value at $p<0.05$. In district Ferozpur, the average number of insecticide applications were 17.86 in IRM villages and 20.70 in non-IRM villages, which was a positive difference of 15.9 per cent and the difference was statistically significant at $p<0.05$ as calculated 't' value 1.99 is higher than table value. In district Mansa, the average number of insecticide applications were 12.92 in IRM villages and 16.67 in non-IRM villages and the difference was 29.02 per cent which was statistically significant at $p<0.01$ as calculated 't' value is 2.87. Overall the difference between IRM and non-IRM villages in Punjab was statistically significant at $p<0.05$ as calculated 't' value is 2.38 which is higher than the table value. The average number of insecticide applications in IRM villages was 13.07 and non-IRM villages were 15.43 with a positive difference of 17.99 per cent. The alternative

Table 5.30: Average Number of Insecticide Applications Against Different Insect Pests of Cotton During 2004

Insect Pest	*Average Number of Insecticide Applications*						*Average of Three Districts*	
	District Bathinda		*District Ferozepur*		*District Mansa*			
	IRM Villages	*Non-IRM Villages*	*IRM Villages*	*Non-IRM Villages*	*IRM Villages*	*Non-IRM Villages*	*IRM Villages*	*Non-IRM Villages*
Bollworm complex	4.50	4.90	12.67	16.95	8.32	11.60	8.50	11.15
Tobacco caterpillar	0.75	0.65	1.85	0.60	2.06	2.45	1.55	1.23
Whitefly	1.30	1.00	2.24	1.80	0.84	0.82	1.46	1.21
Jassid	1.90	2.30	1.00	1.30	1.51	1.80	1.47	1.80
Other pests	0.04	0.00	0.02	0.00	0.10	0.00	0.05	0.00
For growth of cotton crop	0.00	0.00	0.08	0.05	0.08	0.00	0.05	0.02
n	**49**	**20**	**49**	**20**	**49**	**20**	**147**	**60**

Table 5.31: Average Number of Insecticide Applications in Cotton Crop During 2004

Parameter	District Bathinda			District Ferozepur			District Mansa			Average of Three Districts		
	IRM Villages (n=49)	Non-IRM Villages (n=20)	Difference	IRM Villages (n=49)	Non-IRM Villages (n=20)	Difference	IRM Villages (n=49)	Non-IRM Villages (n=20)	Difference	IRM Villages (n=147)	Non-IRM Villages (n=60)	Difference
Average number of insecticides applications 2004[1]	8.45	8.85	(–) 0.40 (4.7)	17.86	20.70	(–) 2.84* (15.9)	12.92	16.67	(–) 3.75** (29.02)	13.07	15.43	(–) 2.351* (17.99)
SD	3.78	2.52	–	6.33	4.93	–	5.44	4.82	–	6.52	6.46	–
(i) Average number of tank mixed insecticide applications in 2004	1.00	1.75	(–) 0.75** (75)	4.98	7.70	(–) 2.720** (54.6)	3.31	5.81	(–) 2.503** (75.6)	3.10	5.10	(–) 2.003** (64.6)
SD	2.05	1.07		4.04	3.13	–	3.18	3.41	–	3.57	3.68	–
(ii) Percentage of farmers using tank mixed insecticides in 2004	33	90	(–) 57	84	95	(–) 11	75	100	(–) 25	64	95	(–) 31
Average number of insecticides application 2004[2]	8.14	6.95	(–) 1.193 (17.2)	13.08	13.10	(–) 0.02 (0.2)	9.67	10.86	(–) 1.19* (12.3)	10.05	10.31	(–) 0.26 (2.6)
SD	5.31	1.50	–	3.45	2.40	–	2.97	1.80	–	13.83	3.18	–
Average number of insecticides applications in Bt cotton	3.60	4.50	(–) 0.9 (25.0)	6.70	4.77	(+) 1.93 (40.5)	4.86	5.33	(–) 0.47 (9.7)	4.63	4.86	(–) 0.23 (4.97)
SD	1.54	1.61	–	3.99	4.58	–	1.53	1.56	–	2.48	2.61	–
Average number of days after sowing for 1st insecticides application	86.2	93	–	86.5	82.0	–	86.78	85.23	–	86.49	86.74	–
SD	10.29	12.48	–	12.00	12.00	–	13.20	8.39	–	12.12	11.24	–

Figures in the parenthesis are the percent difference/change.

1. Mixtures of insecticide have been calculated as two applications; 2. Mixtures of insecticide have been calculated as one application.

*: The difference in average number of insecticide application is statistically significant at $P<0.05$.

**: The difference in average number of insecticide application is statistically significant at $p< 0.01$.

hypothesis that IRM programme will result in less number of average insecticide applications in IRM villages as compared to non-IRM villages is proved.

The difference in average number of tank mixtures applied in IRM villages and non IRM villages was 3.10 and 5.10 respectively which was a difference of 64.6 per cent and the difference was statistically significant at p<0.01 as calculated 't' value was 3.61 which is higher than table value. In all the districts, there was significant difference in the average number of tank mixtures applied in IRM and non-IRM villages at p<0.01. The IRM farmers had applied significantly lesser number of tank mixtures in comparison with non-IRM farmers. The district wise difference in average number of tank mixtures ranged from 54.6 per cent to 75.6 per cent. The percentage of farmers applying tank mixtures was also on lower side in IRM villages than in non-IRM villages. Overall the percentage of farmers using tank mixtures was 64 per cent in IRM villages and 95 per cent in non-IRM villages which was a difference of 31 per cent. In non-IRM villages the five per cent of farmers not applying tank mixtures were mostly those who had sown Bt-cotton. In IRM villages 36 per cent of farmers not using tank mixtures, was the main impact of training.

In case of the tank mixture calculated as one application then there was not statistically significant difference in the average number of insecticide application between the IRM and non-IRM farmers, even though a positive difference of 2.6 per cent, but comparison on that basis would be erroneous. The farmers on an average in both IRM and non IRM villages had applied the first insecticide application 86 days after sowing.

The insecticide use in terms of average number of applications in *Bt* cotton were 4.63 in IRM villages and 4.86 in non-IRM villages which was a difference of 4.97 per cent and was not statistically significant. In Bt cotton, 182.29 per cent less applications were given than in other varieties in IRM villages as the number of average applications were 4.63 in Bt-cotton and 13.07 in other varieties. In non-IRM villages the difference in Bt cotton and other varieties was 217.49 per cent as the average number of insecticide application in Bt cotton were 4.86 and other varieties were 15.43. The reduction in pesticide use as a result of IPM programmes have been reported by Kiss and Meerman (1991), Kenmore (1991), van de Fliert 1993), Sugonyaev (1994), Ragunathan (1995), FAO (1996) and others.

While analysing the data, a trend was observed that farmers' with large landholdings had applied higher number of insecticide applications than small or semi-medium farmers in both IRM and non-IRM villages. The statistical analysis was done to find out the association between operational landholding (hectares) and the average number of insecticide applications by using Pearson's correlation coefficient. There is significant positive correlation between the landholding and the number of insecticide application at p<0.01 (d.f.=205)as the calculated 'r' value is 0.198. As the size of landholding increased so had the number of insecticide applications. The farmers with large holdings are less rational in their decision making about pesticide use. This shows that farmers having large landholding do not have the problem of cash thus more pesticidense. There is no significant correlation between education and pesticide application, but education and number of insecticide applications are negatively correlated. The IRM programme in Punjab resulted in the reduction of the insecticide use in cotton crop in the IRM villages, but the reduction has not been uniform in all the districts. As results indicate (Table 5.31) the farmers of Bathinda district had on an average applied less number of insecticide applications than in district Mansa and district Ferozpur.

5.5.3.6 Insecticide Use: Technical Grade Material

The insecticide use (technical grade material) is another procedure for finding out the reduction in the use of insecticides on account of IPM programmes. Lacewell and Taylor (1980) reported that IPM programmes resulted in reduction in the use of pesticides (technical grade material) in different IPM programmes across the USA. The technical grade material was calculated on the basis of a particular dosage of an insecticide applied and the area of application. On this basis total insecticide use in a district was worked out with respect to area. The data are presented in Table 5.32. The technical grade material of insecticides use as calculated is not the use of insecticide per farmer, but it is the use of insecticides per hectare.

The use of insecticides in IRM villages was 5.602 kg ha^{-1}, whereas in non-IRM villages it was 8.032 kg ha^{-1}, which means a positive difference of 43.40 per cent. The use of insecticides in the IRM villages was 2.430 kg ha^{-1} less than in the non-IRM villages. The difference in district Bathinda between IRM and non-IRM villages was only 5.52 per cent, whereas in Mansa the difference was 28.56 per cent and Ferozpur the difference was 33.55 per cent. In district Ferozpur, the consumption of insecticide is the highest, as the insecticide use was 7.773 kg ha^{-1} in the IRM villages and 10.381 kg ha^{-1} in the non-IRM villages. In district Bathinda, the consumption of insecticide is the lowest, 3.478 kg ha^{-1} in IRM villages and 3.671 kg ha^{-1} in non-IRM villages. One of the reasons for this is that in district Bathinda, maximum area was under *Bt* cotton and even in non-Bt cotton, the average number of insecticide applications was less as compared to other two districts.

The group wise consumption of different insecticides is given in Table 5.32. The consumption of organophosphates group of insecticide is the highest as the recommended dosages of this group are more than other groups which results in higher use and the number of insecticides of this group recommended are also highest. There was a difference of 2.283 kg ha^{-1} in the use of this group in IRM and non-IRM villages. In IRM villages the use of organophosphates was 4.951 kg ha^{-1}, whereas in non-IRM villages it was 7.234 kg ha^{-1}. The use of all the different groups of insecticides (except organochlorinated) was more in non-IRM villages (Table 5.32). The results prove the hypothesis that IRM farmers use of insecticide (technical grade material) was less than non-IRM farmers and the difference is quite high (43.40 per cent).

5.6 Economic Benefits at the Farmers' Level

The economic benefit is the important aspect which influences the adoption or rejection of the technology. Unless there is a clear economic advantage, an IPM programme is not likely to be adopted. To find out the economic benefit of IRM programme, the average input and output of cotton production in IRM villages and non-IRM villages was worked out on the basis of inputs and output of cotton

Table 5.32: Pesticides Use (Technical Grade Material) in Cotton Crop During 2004 (Other Than Seed Treating Chemicals)

Parameter	*District Wise Consumption of Pesticides (kg ha⁻¹)*									*Total of Three Districts (kg ha⁻¹)*		
	District Bathinda			*District Ferozepur*			*District Mansa*					
	IRM Villages	*Non-IRM Villages*	*Diffe-rence*	*IRM Villages*	*Non-IRM Villages*	*Diffe-rence*	*IRM Villages*	*Non-IRM Villages*	*Diffe-rence*	*IRM Villages*	*Non-IRM Villages*	*Diffe-rence*
Insecticide groups												
(i) Synthetic pyrethroids	0.036	0.099	–0.063	0.271	0.535	–0.264	0.145	0.241	–0.096	0.166	0.385	–0.219
(ii) Organophosphates	3.096	3.240	–0.144	6.937	9.426	–2.489	3.712	4.913	–1.201	4.951	7.234	–2.283
(iii) Organochlorinated	0.246	0.203	+0.043	0.466	0.161	+0.305	0.314	0.253	+0.061	0.360	0.187	+0.173
(iv) Carbamates	0.000	0.000	0.000	0.000	0.000	0.000	0.047	0.107	–0.060	0.011	0.020	–0.009
(vi) Nitroguadine	0.049	0.079	–0.030	0.030	0.051	–0.021	0.034	0.054	–0.020	0.037	0.058	–0.021
(vi) Oxadiazine	0.032	0.047	–0.015	0.024	0.077	–0.053	0.083	0.031	+0.052	0.040	0.062	–0.022
(vii) Naturalyte	0.019	0.003	+0.016	0.045	0.132	–0.087	0.048	0.034	+0.014	0.037	0.086	–0.049
Total use of all insecticides (kg ha⁻¹)	3.478	3.671	–0.0193 **(–5.52)**	7.773	10.381	–2.608 **(–33.55)**	4.383	5.634	–1.251 **(–28.56)**	5.602	8.032	–2.430 **(–43.40)**
Total consumption of insecticides (in kg)	1088.623	278.270		3328.357	2163.491		953.952	374.11		5370.932	2815.371	
Herbicides (kg ha⁻¹)	0.426	0.161	+0.265	0.118	0.146	–0.208	0.398	0.118	+0.280	0.282	0.144	+0.138
Total area (ha)	**313**	**75.80**		**428.20**	**208.40**		**217.65**	**66.4**		**958.85**	**350.60**	

Figures in the parentheses are the percentage difference in insecticide consumption (technical grade material) per ha between IRM and Non-IRM villages.

production during 2004. Detailed records of cotton cultivation input and output were made by the respondent farmers in the questionnaires provided to them at the start of cotton growing season in 2004. In Table 5.33, detailed overview and economic analysis of the input and output data of cotton production is given.

5.6.1 Gross Returns

The gross returns were worked out on the basis of seed cotton yield during 2004. The average seed cotton yield varied between 20.13 q ha^{-1} to 26.74 q ha^{-1} in the IRM villages of the three districts. The overall average of seed cotton yield in IRM villages of the three districts was 23 q ha^{-1} with a standard deviation of ±5.00 and in the non-IRM villages was 22.09 q ha^{-1} with a standard deviation of ± 6.38. The variation in seed cotton yield in non-IRM villages was more in comparison to IRM villages. The average sale rate of seed cotton during 2004 varied from district to district and from September 2004 to January 2004. Due to bumper cotton production during 2004, the prices of cotton nose dived, as against an average price of Rs.2550 per quintal in 2003; it was Rs. 1820 per quintal in 2004. It was estimated that there was an overall rise of 30 per cent in cotton production in 2004[1], and farmers had to make a beeline at local markets to sell their produce and that too at lower rates (Figure 5.21). As a result, the farmers did not gain substantially in terms of net returns in 2004 as compared to 2003 (Figure 5.22). Even a good number of farmers did not sell a part of their produce until late January 2005.

In Punjab, the sale rate of seed cotton fluctuated between the months, from Rs. 2180 q^{-1} at the start of first picking to Rs. 1650 q^{-1}. The average sale rate per quintal worked out to be Rs. 1820.24 with a standard deviation of ± Rs. 135.04. Gross returns based on seed cotton rate of Rs. 1820 q^{-1} in case of IRM villages were, Rs. 48667 ha^{-1} district Bathinda, Rs. 40277 ha^{-1} district Mansa and Rs. 36637 ha^{-1} district Ferozpur. In case of non-IRM villages gross returns were, Rs. 47538 ha^{-1} (Bathinda), Rs. 38966 ha^{-1} (Mansa) and Rs. 34107 ha^{-1} (Ferozpur). Overall, gross returns in case of IRM villages were Rs. 41860 ha^{-1} and non-IRM village was Rs. 40204 ha^{-1} (Table 5.33). The difference between gross returns of IRM and non-IRM village was Rs. 1656 ha^{-1} (4.12 per cent), a very small difference.

5.6.2 Cost of Cultivation and Net Returns

The expenditure on cotton production were divided into six categories *viz.* land preparation, sowing of cotton seed, inputs, cost of spraying, hoeing and other expenditures. Land preparation includes pre-sowing irrigation, ploughing and planking. Expenditure on pre-sowing irrigation is computed on the basis average expenditure on pre-sowing as per the information given by the respondent farmers. The expenditure on ploughing and planking is based on the average number of ploughings and plankings multiplied by the uniform market rates. The market rate for one ploughing was Rs. 650 ha^{-1}, and for one planking was Rs. 250 ha^{-1}. Sowing of cotton seed was calculated on the basis of 5 labourers ha^{-1} at the rate of Rs. 70 per labour and consumption of diesel at the rate of 2.5 litres ha^{-1} in tractor drawn seed

[1] "Cotton continues to pile up", *Indian Express*, Dec 13, 2004.

Figure 5.21: Problems of Plenty: Bumper Cotton Crop in 2004, Farmers Making Beeline for Selling their Produce at Local Market

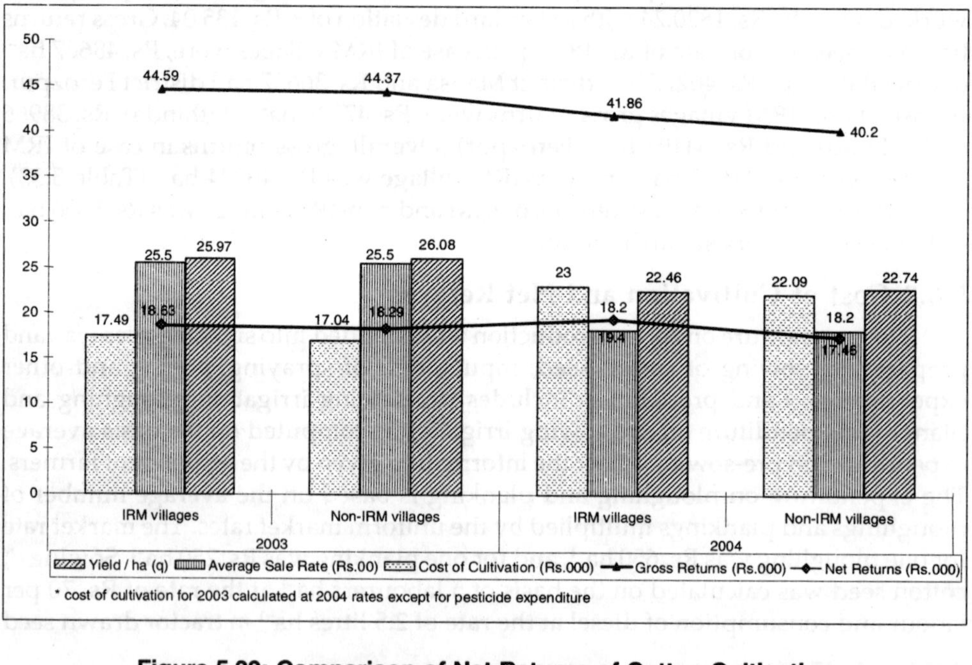

Figure 5.22: Comparison of Net Returns of Cotton Cultivation Between 2003 and 2004

cum fertilizer drill. In case of pre-sowing, on average 16-25 litres of diesel are consumed for irrigating one hectare of land with the pump set.

The expenditure incurred on inputs included seed, chemical fertilizers, farm yard manure, insecticides, herbicides and irrigation. The cost of seed was calculated on the basis of the area under a particular variety of hybrid, non-hybrid, *Bt* cotton, *Desi* cotton, the quantity of seed sown in that area and the cost of a particular variety as reported by the respondent farmer. On this basis the cost of seed per ha in one district was calculated which was divided by the total area under cotton. Same procedure was adopted for calculating the cost of seed ha^{-1} in all the three districts, first separately, then overall for all the three districts. The cost of seed per ha in case of Bathinda is higher as most of the farmers (98 per cent) had cultivated Bt-cotton and the area under Bt-cotton was 57.7 per cent. The average cost of Bt-cotton seed for one hectare was Rs. 5000 as the farmers had purchased Bt-cotton seed from Gujarat and the local dealers at high rates, as the same was not released for Punjab before 2005. The cost of hybrid varieties ranged between Rs. 875 ha^{-1} to Rs. 2250 ha^{-1} depending upon a particular variety. The cost for non-hybrid varieties and *desi* cotton varied between Rs. 312 to Rs. 500 ha^{-1}. In Ferozpur district, the expenditure on cotton seed was less Rs. 967 ha^{-1} (IRM villages) and Rs. 1193 ha^{-1} (non-IRM villages) because the maximum area was under non-hybrid varieties, 80.9 per cent in IRM villages and 77.6 per cent in non-IRM villages.

The cost of chemical fertilizers was calculated at market price and cost of farm yard manure (FYM) was calculated by multiplying average quantity of FYM applied (tons per hectare) with Rs. 3.75 ton^{-1} as the cost of 40 tonne trolley of FYM was Rs. 150. The cost of insecticides and weedicides were also calculated at market price. In case of expenditure on irrigation the average was computed on the basis of expenditure reported by the farmers for irrigating one hectare of land.

The cost of spraying was calculated on the basis of Rs. 100 ha^{-1} per application in case of tractorized and manual sprayer. In case of manual sprayer Rs. 100 were charged for spraying one hectare of area per spray and in case of tractorised sprayer the consumption of diesel varied between 1.5 litre to 2.5 litre ha^{-1} per spray and the labour expenditure. The average number of pesticide applications in a district was multiplied with cost per pesticide application per hactare (IRM and non-IRM villages) to compute spraying expenditure per hectare.

The expenditure on account of hoeing was computed on uniform market rates. The average number of hoeings, tractorised and manual were multiplied with Rs. 250 ha^{-1} and Rs. 625 ha^{-1}, respectively. The other expenditures included cost of picking (Rs. 170 q^{-1}), cleaning of bunds (Rs. 175 ha^{-1}) and cleaning of fields after picking (Rs. 350 ha^{-1}). All these were calculated at uniform market rates. The results from Table 5.33 indicate that total paid out cost per hectare was highest in district Bathinda and lowest in district Ferozpur. But the input- output ratio which is the ratio between the gross returns and cost of cultivation was the highest in district Bathinda (2.04 in IRM villages and 2.02 in non-IRM villages) and lowest in non-IRM villages of district Ferozepur(1.53).The overall input-output ratio in IRM villages was 1.86 and non-IRM villages was 1.77 (Table 5.33) and the difference was

Table 5.33: Average Input and Output of Cotton Production During 2004

Category	District Bathinda		District Ferozepur		District Mansa		Total of Three Districts	
	IRM Villages	Non-IRM Villages	IRM Villages	Non-IRM Villages	IRM Villages	Non-IRM Villages	IRM Villages	Non-IRM Villages
1	2	3	4	5	6	7	8	9
Number of farmers	49	20	49	20	49	20	147	60
Average holding size under cotton during 2004 (ha)	6.26	3.79	8.56	10.42	4.35	3.32	6.39	5.84
Output								
1. Average seed cotton yield (q ha^{-1})	26.74	26.12	20.13	18.74	22.13	21.41	23.00	22.09
Gross returns[1] (Rs ha^{-1}) @ Rs 1820/q	48667	47538	36637	34107	40277	38966	41860	40204
Cost of cultivation (Rs ha^{-1})								
1. Land preparation (Rs ha^{-1})								
(i) Pre-sowing irrigation[7]	505	415	380	363	496	495	460	424
(ii) Ploughing and planking[#]	2308	2406	2293	1875	2055	2126	2219	2103
2. Sowing (Rs ha^{-1})[#]	415	415	415	415	415	415	415	415
3. Inputs								
(i) Seed[2]	2967	2768	967	1193	2387	2377	1942	1757
(ii) Chemical fertilizers[3]	2061	1995	1931	1867	2105	1937	2033	1942
(iii) FYM[4]	188	56	81	24	189	91	141	60
(iv) Insecticides[5]	4576	4927	6985	8714	5907	6978	5793	6875
(v) Herbicides[6]	363	165	124	159	318	121	246	153
(vi) Irrigation[7]	3400	4001	2248	1884	2664	2364	2731	2725

Contd...

Table 5.33–Contd...

Category	District Bathinda		District Ferozepur		District Mansa		Total of Three Districts	
	IRM Villages	Non-IRM Villages	IRM Villages	Non-IRM Villages	IRM Villages	Non-IRM Villages	IRM Villages	Non-IRM Villages
1	2	3	4	5	6	7	8	9
4. Cost of spraying (Rs ha^{-1})[8]								
(i) Insecticides	814	695	1308	1310	967	1086	1005	1031
(ii) Herbicides	88	60	45	25	100	60	78	45
5. Hoeing (Rs ha^{-1})#								
(i) Cost of manual hoeings	750	735	588	490	698	749	679	658
(ii) Cost of tractorized hoeing	350	350	225	300	287	180	287	277
6. Other expenditure (Rs ha^{-1})#								
(i) Picking	4546	4440	3422	3186	3762	3640	3910	3755
(ii) Cleaning bunds	175	175	175	175	175	175	175	175
(iii) Cleaning of cotton field after picking	350	350	350	350	350	350	350	350
Total paid out cost (Rs ha^{-1})	23856	23953	21537	22330	22875	23144	22464	22745
Net income (Rs ha^{-1})	24811	23585	15100	11777	17402	15822	19396	17459
Input-Output ratio	2.04	2.02	1.70	1.53	1.76	1.68	1.86	1.77
Pesticides cost + cost of application: Total production cost (%)	24.80	24.41	39.29	45.71	31.88	35.62	31.70	35.63
Insecticides cost + cost of application: Total production cost (%)	22.59	23.47	38.51	44.89	30.05	34.84	30.26	34.76

#: At uniform market rates.

1. Returns computed using average actual market price; 2. Price of purchased seed or own seed at market price; 3. Fertilizer at market price plus cost of application; 4. FYM @ Rs 150/trolley (40 tonnes); 5–6. Insecticides and weedicides at market price; 7. Irrigation as per expenditure given by a farmer for canal and tube well irrigation; 8. Rs 100/ha.

0.09(5.08 per cent) which was not statistically significant. Overall there was negligible difference in the cost of cultivation of IRM (Rs. 22464) and non-IRM villages (Rs. 22745). The net income in IRM villages was Rs. 19396 ha^{-1} and non-IRM villages was Rs. 17459 ha^{-1} which was 11.09 per cent more in case of IRM villages (Table 5.33).

5.6.3 Pesticide Costs

The total pesticide cost including cost of spraying to total cost of production was 24.48, 39.29 and 31.88 per cent in the IRM villages of districts Bathinda, Ferozpur and Mansa, respectively. Whereas it was 24.41, 45.71 and 35.62 per cent, in the non-IRM villages of districts Bathinda, Ferozpur and Mansa, respectively. Overall the cost of insecticides including spraying cost to total cost of production in the IRM villages was 30.26 per cent and in the non-IRM villages 34.76 per cent a difference of 4.5 per cent. The difference was not more because farmers in non-IRM villages had saved on cost of spraying as they had used more tank mixtures.

The cost of insecticide to total cost of production in Punjab has risen from 2.1 per cent in 1974-75 to 4.6 per cent in 1979-80 to 11.9 per cent in 1984-85 to 15.5 per cent in 1989-90, 13.0 per cent in 1994-95 (Dhaliwal and Arora 2001) and 21.2 per cent in 1998-99 (Sen and Bhatia 2004). The cost of pesticides to total cost of cultivation in case of district Bathinda, was reported to be 50 per cent (Rs. 3666: 7365 $acre^{-1}$) by Shetty (2004), which was termed as pesticide hotspot in Punjab. But the results of this study indicate that Ferozpur district is a pesticide hot spot as the ratio of insecticide cost to total cost of production is 39.29 per cent in IRM villages and 45.71 in non-IRM villages (Table 5.33). This rising trend of the insecticide cost to total cost of production shows the different IPM programmes implemented in Punjab have had no impact in terms of reducing the expenditure on account of insecticide use. A comprehensive study needs to be conducted in the villages covered under IPM programmes in Punjab from time to time to find out the long term impact of these programmes, if any. Such studies for other states of India are also desired. The IPM impact as reflected in project reports of different programmes can be taken as base line data for evaluating the impact of these programmes. This presents a challenging field of research to the scientists from the field of extension education.

5.6.4 Economic Benefits of IRM Programme

The results indicate there was no significant difference in the yield, average number of insecticide applications and average insecticide expenditure between IRM and non-IRM farmers in baseline data of 2003 except in case of pesticide expenditure in district Ferozepur (Table 5.34). The seed cotton yield in IRM villages was 17.49 q ha^{-1} in IRM villages and 17.04 q ha^{-1} in non-IRM villages in 2003. The average insecticide applications was 15.34 (SD±4.40) in IRM villages and 14.93 (SD±3.79) in non-IRM villages and the average insecticide expenditure was Rs. 10305 ha^{-1} (SD±2989) in IRM villages and Rs. 11145 ha^{-1} (SD±3591) in non-IRM villages.

In 2004, there was a shift in the average number of insecticide applications (mixtures taken as one application) in all the IRM villages. In case of district Bathinda, the average number of insecticide applications in 2003 were 14.55 and in 2004 were 8.14 and the mean difference was 6.40 which was significant at $p<0.01$ as calculated

Table 5.34: Economic Benefits of IPM Technology Disseminated Under IRM Project

Parameter	District Bathinda			District Ferozepur			District Mansa			Average of Three Districts		
	IRM Villages (n=49)	Non-IRM Villages (n=20)	Difference	IRM Villages (n=49)	Non-IRM Villages (n=20)	Difference	IRM Villages (n=49)	Non-IRM Villages (n=20)	Difference	IRM Villages (n=147)	Non-IRM Villages (n=60)	Difference
Before the IRM project 2003												
(i) Seed cotton yield[1] (q ha^{-1})	16.01	17.26	–1.25 (7.80)	18.64	17.24	+0.564 (3.27)	17.94	16.61	+1.337 (8.05)	17.49	17.04	+0.449 (2.63)
SD	**3.17**	**3.75**	–	**3.17**	**3.75**	–	**4.06**	**3.20**	–	**3.94**	**3.63**	–
(ii) Average insecticides applications[1] (No.)	14.55	12.95	+1.60 (12.35)	16.48	15.75	+0.727 (4.62)	15.10	16.16	–1.056 (7.00)	15.34	14.93	+0.406 (2.72)
SD	**4.73**	**4.25**	–	**3.96**	**3.37**	–	**4.31**	**2.93**	–	**4.40**	**3.79**	–
(iii) Average insecticides expenditure[1] (Rs.ha^{-1})	10805	10689	+116 (1.08)	10095	12713	–2618*** (25.93)	9994	9975	+19 (0.19)	10305	11145	–840 (8.15)
SD	**3451**	**4109**	–	**2031**	**3846**	–	**3201**	**1924**	–	**2989**	**3591**	–
During the IRM project 2004												
(i) Seed cotton yield[1] (q.ha^{-1})	26.74	26.12	+0.62 (2.37)	20.13	18.74	+1.382* (7.38)	22.13	21.41	+0.72 (3.36)	23.00	22.09	+0.908 (4.11)
SD	**5.26**	**5.68**		**4.52**	**1.74**		**3.97**	**6.09**		**5.00**	**6.38**	
(ii) Average insecticide applications[1] (No.)	8.45	8.85	–0.401 (4.7)	17.86	20.70	–2.84** (15.9)	12.92	16.67	3.75*** (29.02)	13.07	15.43	–2.351** (17.99)
SD	**3.78**	**2.52**	—	**6.33**	**4.93**	–	**5.44**	**4.82**	–	**6.52**	**6.46**	–

Contd...

Table 5.34–Contd...

Parameter	District Bathinda			District Ferozepur			District Mansa			Average of Three Districts		
	IRM Villages (n=49)	*Non-IRM Villages (n=20)*	*Diffe-rence*	*IRM Villages (n=49)*	*Non-IRM Villages (n=20)*	*Diffe-rence*	*IRM Villages (n=49)*	*Non-IRM Villages (n=20)*	*Diffe-rence*	*IRM Villages (n=147)*	*Non-IRM Villages (n=60)*	*Diffe-rence*
(iii) Average number of insecticide applications[1] (Tank mixtures taken as one application)	8.14	6.95	+1.19 (17.2)	13.08	13.10	–0.02 (0.2)	9.67	10.86	–1.19 (12.3)	10.05	10.31	–0.26 (2.6)
SD	**5.31**	**1.50**	–	**3.45**	**2.40**	–	**2.97**	**1.80**	–	**3.83**	**3.18**	–
(iv) Average insecticide expenditure[1] (Rs.ha^{-1})	4576	4927	–351 (7.67)	6895	8714	–1819*** (26.38)	5907	6978	–1070** (18.11)	5793	6875	–1082*** (18.67)
SD	**2364**	**1709**	–	**2258**	**1742**	–	**1960**	**1312**	–	**2385**	**2203**	–

*: Difference significant at p<0.10; **: Difference significant at p<0.05; ***: Difference significant at p < 0.01.

The figures in the parenthesis is per cent change/difference.

For 2003 the actual number of tank mixture of pesticide not known so all sprays taken as one application.

1: The reduction in average insecticides expenditure and average number of insecticide applications (tank mixture taken as one application) during 2004 was significant w.r.t. to 2003 in IRM and non-IRM villages at p<0.01. The increase in yield/ha was significant in comparison to 2003 at p<0.01.

't' value was 6.387. In district Ferozpur, the average insecticide applications during 2003 were 16.48 and 13.08 with a mean difference of 3.40 which was significant at $p<0.01$ as calculated 't' value was 5.85. In case of district Mansa, the average number of insecticide applications during 2003 were 15.10 and during 2004 were 9.67 with a mean difference of 5.43 which was significant at $p<0.01$ as calculated 't' value was 8.16. The reductions in average number of insecticide applications were significant in case of non-IRM villages also. In 2003 the average number of insecticide application were 14.93 and in 2004 were 10.31 with a mean difference of 4.63 which was significant at $p<0.01$ as calculated 't' value was 8.62. The results show that the trend in reduction of insecticides use were in both the IRM and non-IRM villages as the farmers reported that insect pest problem of America bollworm was less in 2004 than in 2002 and 2003.

The reduction in the expenditure of insecticides in 2004 when compared to 2003 was significant in the IRM and non-IRM villages of the three districts. During 2003 the average expenditure on insecticide use in the IRM villages was Rs. 10305 ha^{-1} and non IRM villages was Rs. 11145 ha^{-1}. In 2004 the average expenditure on insecticides was 5793 ha^{-1} in the IRM villages and non-IRM villages was 6875 ha^{-1} (Table 5.34). The reduction of Rs. 4512 ha^{-1} in case of the IRM villages which was significant at $p<0.01$ is calculated 't' value was 15.92 and the reduction of Rs. 4270 ha^{-1} in case of the non-IRM villages was also significant at $p<0.01$ as calculated 't' value was 8.72. There was overall increase in yield per unit area in 2004 in the IRM and non-IRM villages, when compared with 2003. The average yield per hectare in the IRM villages in 2003 was 17.49 q and non-IRM villages was 17.04 q. During the year 2004 the average yield increased by 5.51 q ha^{-1} in the IRM villages which was significant at $p<0.01$ as calculated 't' value was 10.04 and in the non-IRM villages it increased by 5.05 q ha^{-1} which was significant at $p<0.01$ as calculated 't' value was 6.83. The district wise detail about pesticide expenditure and yield in 2004 and 2003 is given in Table 5.34. The results reveal that the reduction in insecticide expenditure and increase in yield had taken place in the IRM as well as non-IRM villages. On the basis of this data, inferences regarding the economic benefits of IRM programme can not be made.

The data were compared between the IRM and non-IRM villages for 2004 as was done for 2003. Baseline data of 2003 did not show any significant difference between the IRM and non-IRM villages, but the data of 2004 showed significant difference in average insecticide applications in district Ferozpur and district Mansa, but no significant difference was observed in district Bathinda. The average insecticide expenditure in the IRM villages during 2004 was Rs. 5793 ha^{-1} and in the non-IRM villages was Rs. 6875 ha^{-1}. The farmers in the IRM villages had on an average spent Rs. 1082 ha^{-1} lesser than non-IRM farmers which was significant at $p<0.01$ as calculated 't' value was 3.15. The IRM programme had resulted in significant difference in insecticide expenditure between the IRM and non-IRM farmers. If extrapolated for the sample of 147 IRM farmers, IRM farmers had saved Rs. 10, 37, 475.70 on an area of 958.58 hectares. This is the economic benefit of IPM practices disseminated under the IRM programme. Yield difference between the IRM and non-IRM farmers in 2004 was 4.11 per cent. The average seed cotton yield in the IRM villages was

23.00 q ha^{-1} and in the non-IRM villages was 22.09 q ha^{-1}. The difference was not significant. The district wise data is given in Table 5.34.

The results indicate that the IRM farmers benefited in terms of saving on insecticide expenditure without adversely affecting the yield. The IPM programmes around the world have shown the economic benefits in terms of reduced pesticide expenditure (Lacewell and Taylor 1980, NRC 1989, van de Fliert 1993, Peshin and Kalra 1998, Russell 2004 and others). But the challenge is that the information regarding results of these economic benefits need to be passed on to the participating farmers in IPM programmes, which has not been the case with IPM programmes. The farmers must be taught to make simple farm economic analysis to know the economic benefits.

On an average $ 5 (Rs. 230) are reported to be spent on each farmer in IRM programme (Russell 2004), that implies that on 150 sample respondents in IRM villages Rs. 34500 were spent, which gave a return of Rs. 1037475 in terms of saving on insecticides.

5.6.5 Inferences

The results indicate that there was hardly any difference in the adoption of resistant varieties, timely sowing, weed management, fertilizer dosages and other cultural practices between IRM and non-IRM farmers. The significant difference was observed in the adoption of chemical control practices and the expenditure incurred on insecticides. IRM farmers had applied significantly lesser number of insecticide applications and incurred significantly less expenditure on insecticides, even though there was no significant increase in yields. The IRM farmers had applied significantly lesser number of tank mixtures of insecticides, and used to observe insect pest problem before spraying insecticides. The overall picture of the benefits of the IRM programme in terms of reduction in insecticide applications, insecticide use (technical grade material), insecticide expenditure and comparison of seed cotton yield, with/without and before/during IRM programme are given in Figures 5.23, 5.24, 5.25 and 5.26. The achievement of IRM has been limited to proper insecticide usage, which resulted in

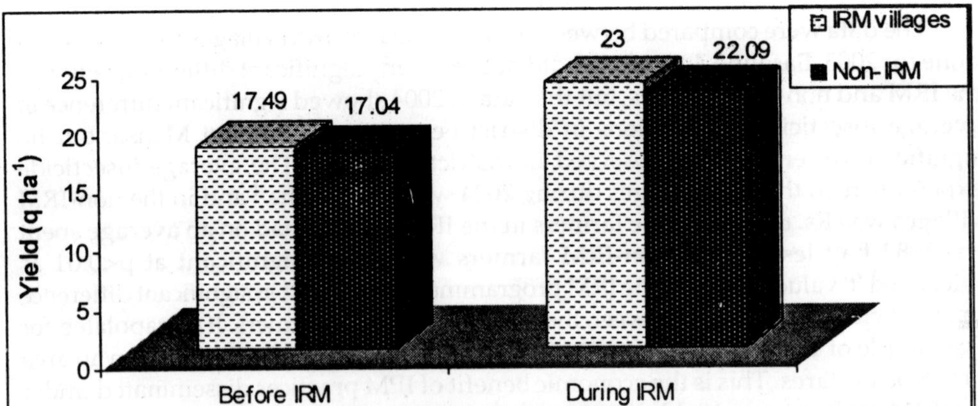

Figure 5.23: Seed Cotton Yield Before/During and With/Without IRM Programme

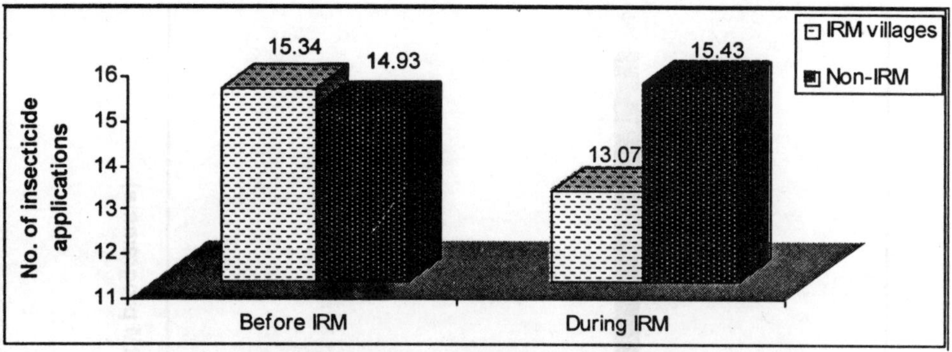

Figure 5.24: Insecticide Applications Before/During and With/Without IRM Programme

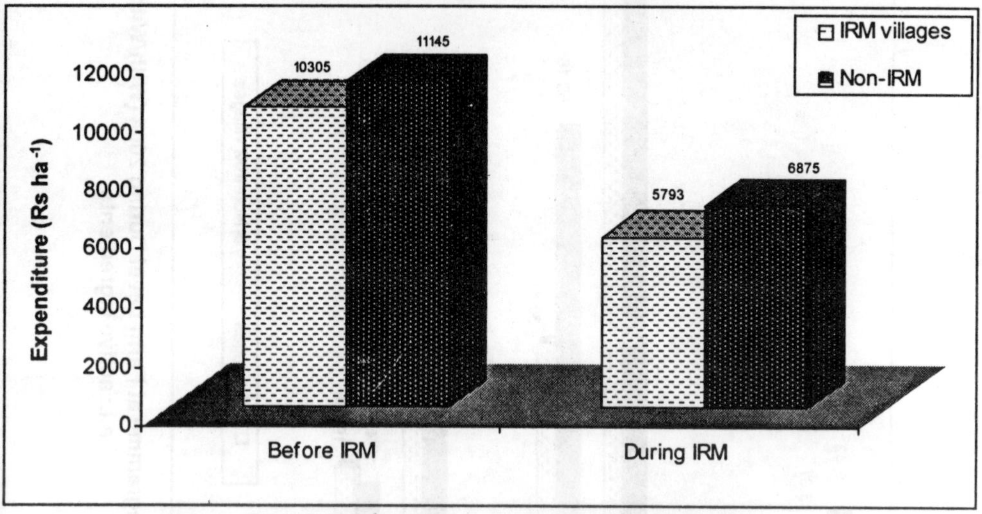

Figure 5.25: Insecticide Expenditure Before/During and With/Without IRM Programme

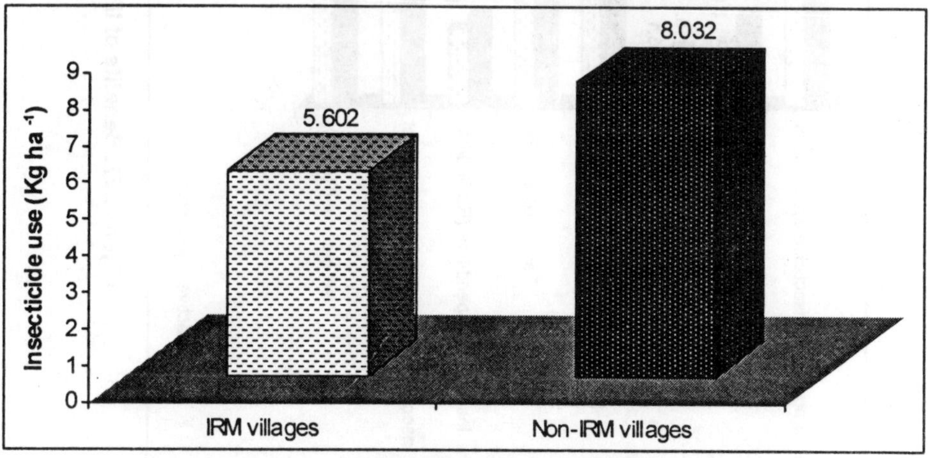

Figure 5.26: Insecticide Use (Technical Grade Material) With/Without IRM Programme

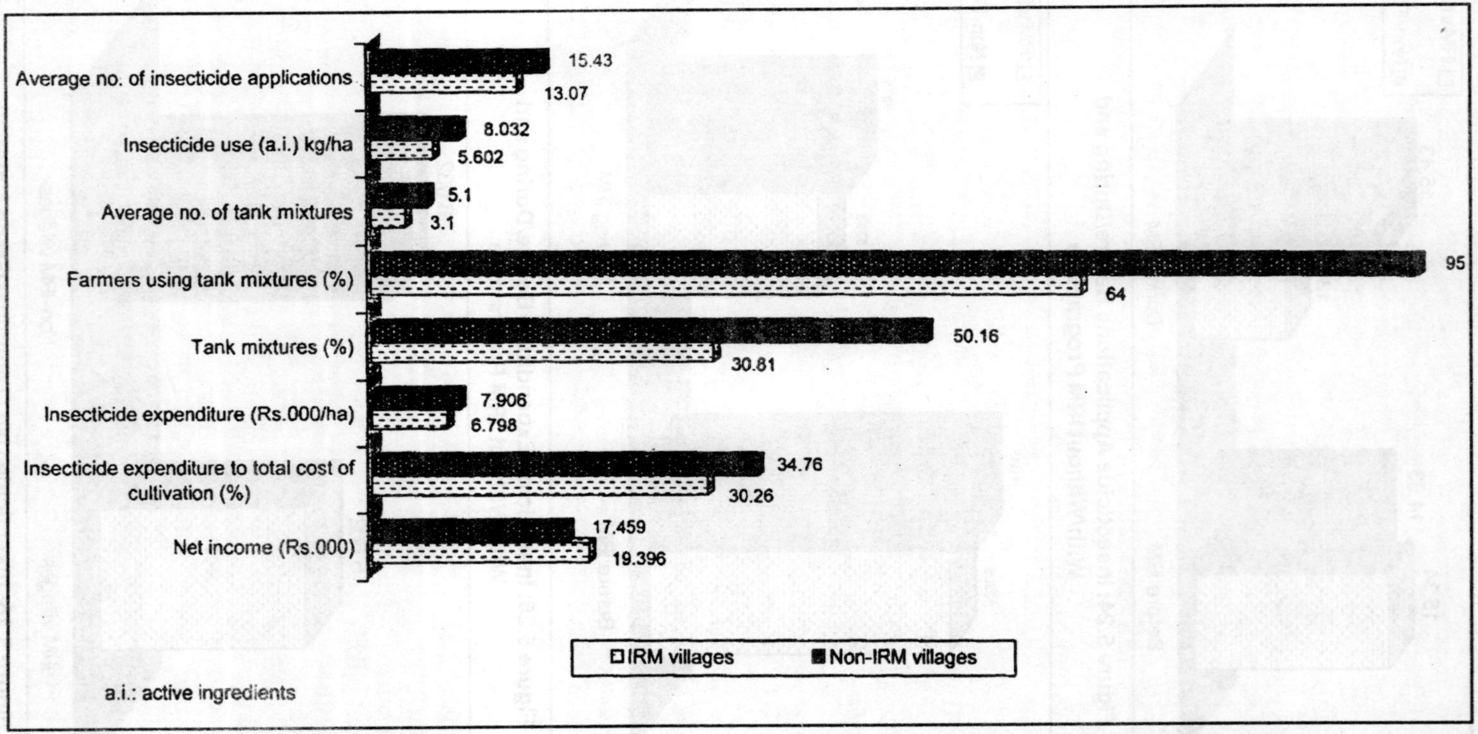

Figure 5.27: Benefits of IRM Programme at Farm Level During 2004 (With/Without IRM programme)
a.i.: active ingredients

reduction in the insecticide use, average number of insecticide application, average number of tank mixtures and average insecticide expenditure (Figure 5.27). The other aspects which include cultural practices need to be advocated by starting regular farmers training before the start of sowing.

5.7 Adoptability and Constraints

Rogers (1983, 1995) recognized five attributes of an innovation affecting the adoption, these are relative advantage, compatibility, observability, complexity and trialability, which in turn affect the rate of adoption and, in turn, the rate of diffusion. Many adoption studies have shown the importance of these aspects (Fliegel and Kivlin 1962, Fliegel *et al.*, 1967). The adoption of technology for natural resource management and conservation, such as soil conservation, IPM, irrigation management, are considered apart, from the use of conventional green revolution inputs, such as high yielding varieties, fertilizers and pesticides (Caswell *et al.*, 2000). In comparison with use of single measure such as pesticides, IPM appears, and often is, complex, its effect is rarely immediately observable (Dent, 1995).

The study intended to analyse how the attributes of innovation effect the adoption of selected IPM practices like timely sowing of cotton crop, adoption of PAU recommended resistant varieties, seed treatment, use of ETL for insecticides application and IRM strategy for insecticide use. The IPM practices being not a single measure to work out the adoptability of IPM in sum total (together), but adoptability of each practice was separately calculated to predict the adoption of these practices in coming year. The adoptability and constraints of IPM practices were studied in the IRM villages only.

5.7.1 Attributes of Timely Sowing of Cotton Crop

The respondent IRM farmers were asked to list the reasons for the adoption of April sowing of cotton crop. The reasons for April sowing of cotton crop were higher seed cotton yield (86 per cent), timely wheat sowing (61 per cent), less pest damage (85 per cent), and early crop maturity (47 per cent). These reasons were also referred to as relative advantages of timely sowing of cotton crop by the respondents. About ten reasons/relative advantages of April sowing of cotton crop were reported by the farmers. But in case of open ended questions the main three reasons referred to were: less pest attack, higher seed cotton yield and timely wheat sowing. The other reasons/ relative advantages referred to are given in Table 5.35. One per cent of the IRM farmers reported that with the change in pest scenario, timely sown cotton crop has more American bollworm infestation.

The other attributes of an innovation that is compatibility greatly influences the timely sowing of cotton crop in Punjab. Even though, the farmers referred it to be compatible with cotton-wheat crop rotation (44 per cent) but 47 per cent referred it to be partially compatible. The IRM respondents reported that it is compatible with small size of land holding as the farmers are able to harvest wheat timely and are able to complete the sowing in the month of April, but it is not compatible with large land holding (27 per cent farmers). One of the most important factors, which affect its compatibility, is lack of availability of canal irrigation water as farmers are not able to complete pre-sowing irrigation timely, and the ground water is not fit for irrigation due to its poor quality and farmers avoid its use for pre-sowing irrigation. The percentage of farmers reporting that timely sowing not compatible with irrigation facilities was 77 per cent.

Table 5.35: Attributes of Timely (April) Sowing of Cotton Crop

Attribute	*District Wise Percentage of Farmer*			*Overall Percentage of Three districts*
	Bathinda	*Ferozepur*	*Mansa*	
Relative advantages				
(i) Higher cotton yield	77	92	90	86
(ii) Timely wheat sowing	58	61	63	61
(iii) Higher wheat yield	6	10	28	15
(iv) Less pest damage	86	92	78	85
Less pest complex	67	80	44	64
Less ABW	19	12	33	21
(v) Less number of pesticides applications required	15	10	6	10
(vi) Early crop maturity and picking	42	37	63	47
(vii) Timely crop rotations	0	4	0	1
(viii) No damage due to summer heat at crop establishment stage	15	8	14	12
(ix) Less weeds	4	0	4	3
(x) Overall cost of cultivation less	4	10	4	6
Disadvantages				
(i) More ABW attack	2	2	0	1
(ii) Less seed cotton yield in 2004	4	0	2	2
Compatibility				
(i) Fully with cotton-wheat crop rotation	40	27	65	44
(ii) Partially with cotton-wheat crop rotation	50	55	35	47
(iii) Cotton-mustard crop rotation	6	33	2	14
(iv) Small size of land holding	25	16	24	22
(v) Irrigation facilities	2	2	8	5
Non-Compatibility				
(i) Preceding wheat crop	19	24	14	19
(ii) Size of land holding	19	51	10	27
(iii) Irrigation facilities	96	65	63	77
n	48	49	49	146

Contd...

Table 5.35–Contd...

Attribute	District Wise Percentage of Farmer			Overall Percentage of Three districts
	Bathinda	Ferozepur	Mansa	
Observability of benefits				
(i) Higher cotton yield	89	90	91	91
(ii) Less pest damage	89	90	79	86
(iii) Timely wheat sowing/ higher yield	11	0	0	4
(iv) Good cotton crop	4	2	0	2
n	**46**	**49**	**47**	**142**

Multiple responses.

Decimals rounded up to nearest whole numbers.

Higher yield of seed cotton quantified by five farmers, April 30-35q ha^{-1}, Early May 15-20q ha^{-1}, Late May 7.5-10q ha^{-1}, Average higher yield 5-9q ha^{-1}, ABW:American bollworm.

Observability of benefits of April sown cotton crop as reported by respondent IRM farmers were, higher seed cotton yield (91 per cent), less pest damage (86 per cent) and other observable benefits as listed in table 5.35. Five farmers even quantified the higher seed cotton yield, which ranged between 5-9 q ha^{-1}.

5.7.2 Constraints in Timely Sowing of Cotton Crop

The IRM farmers who had either sown cotton crop in April as well as May, or had sown only in May or later, reported that shortage of canal irrigation is the most important constraint in timely sowing of cotton crop. IRM farmers who had partially sown in April, 79 per cent of them reported shortage and non-availability of canal water as the limiting factor in completing the sowing in April, and 38 per cent, reported that poor quality of ground water is another important constraint (Table 5.36). The incomplete/late harvesting of wheat also limits the complete sowing of cotton crop in April. The IRM farmers who had sown whole area under cotton crop late also reported the irrigation water being the limiting factor. The different constraints as reported by IRM farmers are listed in Table 5.36.

A farmer in village Malwala, district Bathinda has solved the problem of shortage of canal water for irrigation, by applying the pre-sowing irrigation in the standing wheat crop some days before harvesting of wheat crop, and on harvesting of wheat crop immediately goes for land preparation and completes sowing in the month of April, even though his total land holding is 25 hectares and had cultivated cotton on 10.4 hectares.

5.7.3 Attributes of Recommended Resistant Varieties

The varieties recommended by the PAU, Ludhiana are not widely adopted as discussed in this chapter under section 5.5. A small number of the IRM farmers responded to this question as others expressed ignorance about the attributes of these varieties. The recommended resistant varieties about which response of farmers was sought were Ankur 651, White gold, LHH 144, F1861 and LH1556. In district Mansa, only two varieties *viz.* Ankur 651and White gold were reported by the farmers.

Table 5.36: Constraints in the Adoption of Timely Sowing

Constraint	*District Wise Percentage of Farmer*			*Overall Percentage of Three districts*
	Bathinda	*Ferozepur*	*Mansa*	
Partially adopting timely sowing				
(i) Shortage/non-availability of canal water	88	80	71	79
(ii) Poor quality (No. 2/No. 3) of tube well water	25	33	56	38
(iii) Late/incomplete harvesting of wheat crop	25	4	10	13
(iv) Delayed land preparation	18	0	2	6
(v) More land holding/not possible to complete sowing in April	8	0	7	5
(vi) Timely sowing Mustard + cotton crop rotation	0	7	0	2
n	40	45	41	126
Non-adoption of timely sowing				
(i) Non availability of canal water	50	100	50	60
(ii) Late harvesting of wheat crop	50	100	63	65
(iii) Not owning a tractor	2	0	13	5
n	8	4	8	20

Multiple response.

Decimals rounded up to nearest whole numbers.

Some farmers give pre-sowing irrigation in standing wheat crop.

5.7.3.1 Relative Advantages

The results from table 5.37 indicate that Ankur 651 is the most popular among the PAU recommended varieties as it was reported to be having relative advantage in terms of, less insect pest attack (45 per cent farmers), high yielding (63 per cent farmers), early fruiting (16 per cent farmers), and resistant to CLCuV (12 per cent farmers). The relative advantages of White gold were referred to by two to 12 per cent of IRM farmers and relative advantage of LHH144, F1861 and LH1556 were referred to by a very small percentage of farmers (Table 5.37). In all the districts Ankur 651 is a preferred variety in terms of less insect pest attack and higher yield.

Table 5.37: Attributes of PAU Recommended Resistant/Tolerant Varieties

Attribute	District Wise Percentage of Farmers												Overall Percentage of District				
	District Bathinda					District Ferozepur					District Mansa						
	Ankur 651	White Gold	LHH 144	F 1861	LH 1556	Ankur 651	White Gold	LHH 144	F 1861	LH 1556	Ankur 651	White Gold	Ankur 651	White Gold	LHH 144	F 1861	LH 1556
Relative advantage																	
Less insect-pest attack	33	7	11	4	0	20	7	7	27	0	67	10	45	10	8	10	0
Less CLCV infestation	4	0	4	0	0	7	0	0	13	7	14	5	12	2	2	4	2
High yielding	30	7	7	0	4	47	0	0	7	7	81	19	63	12	4	2	2
Less pesticide use	11	0	4	0	0	0	0	0	0	7	10	5	10	2	2	0	2
Early fruiting/maturity	4	0	0	0	0	20	0	0	0	0	19	0	16	0	0	0	0
Late sown	0	0	0	0	0	0	0	0	0	13	0	0	0	0	0	0	4
n			**27**					**15**			**21**				**51**		
Observability																	
Less insect-pest attack	47	20	7	7	7	20	7	0	13	0	71	4	53	9	2	5	2
Less CLCV infestation	7	0	0	0	0	7	0	0	13	7	4	0	5	0	0	4	2
High yielding	47	20	7	0	7	0	0	0	7	7	75	17	59	13	2	2	2
Less pesticide use	7	0	0	0	0	0	0	0	0	0	13	0	9	0	0	0	0
Big fruiting body	0	0	0	0	0	0	0	0	7	0	0	0	0	0	0	4	0
Early maturity	7	0	0	0	0	20	0	0	0	0	0	0	13	0	0	0	0
n			**15**					**15**			**24**				**54**		

Contd...

Table 5.37–Contd...

Attribute	District Wise Percentage of Farmers												Overall Percentage of District				
	District Bathinda					District Ferozepur					District Mansa						
	Ankur 651	White Gold	LHH 144	F 1861	LH 1556	Ankur 651	White Gold	LHH 144	F 1861	LH 1556	Ankur 651	White Gold	Ankur 651	White Gold	LHH 144	F 1861	LH 1556
Compatibility																	
Compatible in the area	37	7	33	11	4	28	4	4	40	20	56	19	41	10	13	16	8
Compatible with irrigation facilities	11	0	4	4	0	4	0	0	12	0	7	0	8	0	1	5	0
Compatible with picking	0	0	0	0	0	0	0	0	40	0	0	19	0	6	0	13	0
n			**27**					**25**			**27**				**79**		
Not compatible																	
In the particular area	53	7	7	7	0	33	0	0	0	0	53	12	48	7	2	2	0
Water availability for irrigation	0	0	0	0	0	8	0	0	0	0	0	0	2	0	0	0	0
Picking	73	0	13	0	0	25	0	0	0	0	100	0	70	0	5	0	0
Soil of the area	7	0	0	0	0	0	0	0	0	0	0	0	2	0	0	0	0
n			**15**					**12**			**17**				**44**		

Decimals rounded up to nearest whole number.

CLCuV: Cotton leaf curl virus.

5.7.3.2 Observability of Benefits

The observability of benefits were mostly reported in case of Ankur 651, as 53 per cent of the IRM farmers who responded to this question reported it to exhibit less insect pest attack, especially resistant to jassid and 59 per cent of the 54 IRM respondents reported it to be high yielding.

5.7.3.3 Compatibility

Out of the total sample of 147 IRM farmers, 79 reported that one of the other recommended varieties are compatible in their area. Out of the 79 IRM farmers 41 per cent reported that Ankur 651 is compatible in their area, but none of the farmers reported it to be compatible with picking and 44 IRM farmers reported that recommended resistant varieties are not compatible in their area. Ankur 651 was reported to be not compatible by 48 per cent of the 44 IRM farmers, as far as the soil type of their fields is concerned, as it thrives well in heavy and fertile soils as reported by them. The most important reason for Ankur 651 being not compatible is it small boll, small loculi and as such not compatible for picking as reported by 70 per cent of 44 IRM farmers.

5.7.3.4 Reasons for Cultivating Recommended Resistant Varieties

The reasons given by the farmers for cultivating the PAU recommended resistant varieties were, high yielding; early maturity and for trial purpose in case of Ankur 651. In district Mansa, three IRM farmers reported that Ankur 651 was sown as Ankur 2226 was in short supply during 2004, and rated the later as better variety than former. IRM farmers had also sown a particular variety on the advise of other farmers. The number of IRM farmers cultivating recommended resistant varieties was very less (Table 5.38).

5.7.3.5 Constraints in the Adoption of Recommended Resistant Varieties

The extent of adoption of recommended resistant varieties being very low, it was logical to find out the constraints in the adoption of these varieties. The results from Table 5.39 indicate that IRM farmers reported different limiting factors for the adoption of the recommended varieties.

Ankur 651, even though most popular of the PAU recommended hybrid varieties is not compatible with respect to its small boll size, small loculi, which results in problems at the time of picking. These were the main limiting factors in its adoption (Table 5.39). The other constraints were: drying of the plant at maturity and not being fit in sandy soils. Another variety of the same seed company Ankur 2226 was reported to be a better variety than Ankur 651.

White gold is not one of the preferred varieties, even the awareness about this variety was low. Eight per cent of the IRM farmers, reported it not being a good variety. The PAU developed resistant varieties namely LHH 144 and FL1861were reported to be less yielding by five per cent of the IRM farmers. The LHH-144 the only hybrid variety developed by PAU, was reported to be not good for picking as it injures the hands while picking

Table 5.38: Reasons for Sowing PAU Recommended Resistant Varieties

Reason/Relative Advantage	District Wise Number of Farmers										
	District Bathinda					District Ferozepur			District Mansa		
	LHH 144	Ankur 651	White Gold	F 1861	LH 1556	Ankur 651	F 1861	LH 1556	Ankur 651	White Gold	F 1861
Good variety	0	0	0	1	0	0	1	0	0	0	0
Less Jassid/pest damage	0	1	1	1	2	0	0	0	2	0	0
Higher yielding	0	4	2	0	0	3	2	0	9	3	0
Less irrigation required	0	0	0	0	0	0	2	2	0	3	0
Early maturity	0	2	1	0	0	3	0	0	6	0	0
Bold fruiting body	0	0	0	0	0	0	1	0	0	0	0
PAU recommended variety	0	0	1	1	0	0	2	0	1	0	0
For trial purpose	0	1	1	1	0	3	2	4	0	0	0
Advice of other farmers	1	0	0	0	0	0	1	0	0	0	0
Timely wheat	0	0	0	0	0	1	0	0	1	0	0
Late sown variety	0	0	0	0	0	0	0	0	0	0	0
Ankur 2226 not available	0	0	0	0	0	0	0	0	3	0	0
Previously tried variety	0	0	0	0	0	3	2	6	4	3	1
n	1	4	2	1	2	9	7	6	12	4	1

Table 5.39: Constraints in the Adoption of PAU Recommended Resistant Varieties (Reasons other than constraints also included), as per cent IRM Farmers Reporting

Constraint	*District Wise Percentage of Farmer*			*Overall Percentage of Three districts*
	Bathinda (n=47)	*Ferozepur (n=46)*	*Mansa (n=49)*	*(n=142)*
Ankur 651				
(i) Less yielding than Bt cotton	11	0	0	4
(ii) More bollworm damage	4	0	0	1
(iii) Small boll size	19	4	43	23
(iv) Loculi not good (small)	6	4	12	8
(v) Not good for picking	11	4	10	8
(vi) Dries up at maturity	6	4	2	4
(vii) Not fit for sandy soil	0	0	10	4
(viii) Ankur 2226 better variety	0	0	18	6
White Gold				
(i) Not good variety, late fruiting	9	2	14	8
LHH 144				
(i) Less yielding	9	2	4	5
(ii) Late maturity	6	2	0	3
(iii) More pest attack	4	2	2	3
(iv) Less height	2	2	0	1
(v) Injures while picking	0	15	4	6
F 1861				
(i) Low yielding	4	11	0	5
(ii) More bollworm attack	4	2	0	2
(iii) No synchronized fruiting	0	4	0	1
LH 1556				
(i) More bollworm attack	4	0	0	1
(ii) F 414 (Jhurar) better variety	0	9	0	2
Other constraints				
(i) No previous knowledge/ experience of resistantvarieties recommended by PAU	21	48	22	30
(ii) Authentic PAU seed not available	15	9	0	8
(iii) No good variety of cotton given by PAU	4	0	6	4
(v) Rasi 134 better than all PAU varieties	4	0	0	1

Contd...

Table 5.39–Contd...

Constraint	District Wise Percentage of Farmer			Overall Percentage of Three districts (n=142)
	Bathinda (n=47)	Ferozepur (n=46)	Mansa (n=49)	
(v) *Bt* cotton resistant than all other varieties	13	7	2	7
(vi) No new variety given by PAU	0	0	6	2
(vii) No knowledge of PAU recommended hybrid-varieties	0	0	11	4
(viii) Cotton cultivated after 1995-96	0	9	0	3
(ix) Shortage of water so hybrid-varieties not sown	0	22	0	7

Multiple responses.

Decimals rounded up to nearest whole numbers.

The other constraints reported by IRM farmers were: no previous knowledge/experience (30 per cent), authentic seed of PAU varieties not being available (8 per cent), no good cotton variety developed by the PAU (4 per cent), no knowledge of the PAU recommended hybrid varieties (4 per cent) and *Bt* cotton being a better variety (7 per cent).

5.7.3.6 Reasons for Cultivating Other Varieties

The maximum area in IRM villages was under the varieties which are not bracketed under resistant varieties by the PAU. As already discussed that, under the IRM programme, IRM farmers were asked either to cultivate PAU recommended resistant varieties or the varieties farmers had previously tested and found compatible in their area provided, the variety was short duration, resistant to CLCuV and resistant to early season sucking pests. The different varieties cultivated by farmers are given in Appendix I.

The reasons for cultivation of these varieties were sought on open ended questions, so that farmers would be free to express their view points. The results from Table 5.40, indicate that majority of the IRM farmers (54 per cent) reported that the varieties they had cultivated during 2004 were good/higher yielding than those recommend by the PAU. The varieties like Ankur 2226 and some of the other varieties were relatively less prone to pest damage, as reported by 19 per cent of the IRM farmers. The *Bt* cotton which was cultivated by majority of the IRM farmers in district Bathinda and district Mansa was reported to be resistant to bollworm by 24 per cent of the respondents. The respondent IRM farmers also reported that the varieties they cultivated required less pesticides applications (14 per cent). The varieties like F1378 and Rawat Saria were cultivated because the fruiting in these varieties is synchronized as reported by five per cent of the IRM farmers. Some of the varieties like F1378 were cultivated in those areas, where there was shortage of irrigation water, as reported by 16 percent of IRM farmers.

Table 5.40: Relative Advantages/Reasons for Sowing Different Varieties of Cotton other than Recommended Resistant Varieties

Relative Advantage/ Reason	*District Wise Percentage of Farmers*			*Overall Percentage of Three districts (n=146)*
	Bathinda (n=48)	*Ferozepur (n=49)*	*Mansa (n=49)*	
Good/higher yield	65	37	61	54
Less pest damage/attack	40	8	10	19
Resistant to bollworms (*Bt* cotton)	35	14	22	24
Less pesticide applications required	12	10	18	14
Synchronized fruiting	4 (F 1378)	10 (Rawat Saria)	2 (F 1378)	5
Less irrigation water requirement	10 (F 1378)	29	10 (F 1378)	16
Late sown variety	2 (F 1378)	6	6 (F 1378)	5
Good fibre quality (F 414)	0	8	0	5
Early fruiting/big boll and loculi size/early maturity	6	10	31	16
For trail in 2004	35	59	43	46
Results of previous year trial	10	10	4	8
Results/advice of other farmers/dealers	10	6	8	8
Previous experience	33	47	31	37
Other reasons	–	18	14	11

Multiple responses.

Reasons for Bt cultivation: Higher yield (16 per cent), Resistant (12 per cent), Less/no BW (16 per cent) less pesticide applications required (10 per cent), Trial (12 per cent), Advice of others (4 per cent) Previous experience (30 per cent)

The varieties having good fibre quality was reported by five per cent IRM farmers, as the reason for adoption. The other reasons were, early fruiting/big boll and loculi size, and early maturity as was reported by 16 per cent of the IRM farmers. Forty-six per cent of farmers had tried a number of varieties for trial purpose, which goes on to show that farmers also experiment of their own, which is complimented by the fact that 37 per cent of the IRM farmers had adopted those varieties which were tried by them previously (Table 5.40).

5.7.3.7 Important Attributes for Adoption of a Variety

The respondent IRM farmers were asked to rank in order, the three important characteristics of a variety which influences their decision to adopt a variety. The results are given in the Table 5.41.

Table 5.41: Important Attributes for Adoption of a Particular Cotton Variety as Ranked by IRM Farmers

Rank/Attribute	*District Wise Percentage of Farmers*			*Overall Percentage of Three districts (n=146)*
	Bathinda (n=48)	*Ferozepur (n=49)*	*Mansa (n=49)*	
Rank I				
(i) Higher yielding	54	69	49	58
(ii) Resistant variety/less pesticide applications	27	22	33	27
(iii) Authentic seed	6	0	8	5
(iv) Early maturity	6	2	6	5
(v) Less water requiring	0	8	0	3
(vi) Good loculi size/lint quality	0	2	2	1
Rank II				
(i) Higher yielding	33	24	31	29
(ii) Resistant variety/less pesticide applications	38	31	29	32
(iii) Authentic seed	4	4	0	3
(iv) Early maturity	4	4	18	9
(v) Less water requiring	0	10	0	3
(vi) Good loculi size/lint quality	10	12	16	13
Rank III				
(i) Higher yielding	6	4	14	8
(ii) Resistant variety/less pesticide applications	10	10	12	7
(iii) Authentic seed	4	2	8	5
(iv) Early maturity	2	4	10	5
(v) Compatible with wheat crop	2	0	0	1
(vi) Less water requiring	0	6	4	3

Decimals rounded up to nearest whole numbers

Rank one was given to higher yielding by 58 per cent, resistant varieties by 27 per cent and authentic seed/early maturity by five percent of the IRM farmers. Rank two was given to higher yielding by 29 per cent, resistant varieties by 32 per cent, good loculi size/lint quality by 13 per cent of the IRM farmers. Rank three was given to higher yielding by eight per cent, resistant varieties seven per cent, and authentic seed and early maturity by five per cent of the IRM farmers.

5.7.4 Adoptability of Seed Treatment

Seed treatment, which is recommended to avoid the infestation of diseases and to delay the onset of Jassid is not a popular practice with the IRM farmers. The hybrid varieties of cotton are mostly treated by the seed companies, as reported by the IRM farmers. The reason of using treated seed was that seed is already treated, as reported by 94 per cent of the IRM farmers (Table 5.42). Only 25 per cent, of the IRM farmers reported that seed treated with gaucho 70 WS delays the attack of jassid and six per cent reported that treated seed reduces the chances of CLCuV. Six per cent, of the IRM farmers reported, that there is less termite attack in treated seeds.

Table 5.42: Adoption of Seed Treatment/Treated Seed and Reasons for its Adoption

Practice/Reason	*District Wise Percentage of Farmers*			*Overall Percentage of Three districts*
	Bathinda	*Ferozepur*	*Mansa*	
Seed treatment/treated seed	90	45	92	75
(i) Seed treatment#	13	10	2	8
(ii) Treated seed#	90	35	92	72
No seed treatment/treated seed	10	55	8	25
n	**48**	**49**	**49**	**146**
Reasons#				
(i) As seed already treated	95	77	100	94
(ii) Delayed attack of jassid	40	23	11	25
(iii) Good germination	2	0	4	3
(iv) Less disease infestation/ CLCuV	14	0	2	6
(v) Less termite damage	5	9	6	6
(vi) Given by dept. of Agric. for trial	0	4	0	1
n	**43**	**22**	**45**	**110**

#: Multiple responses. Decimals rounded up to nearest whole numbers.

5.7.4.1 Attributes of Seed Treatment/Treated Seeds

The different attributes of treated seed as reported by the IRM farmers were in terms of relative advantage, observability and complexity of seed treatment. The relative advantages of treated seed were in terms of less/delayed jassid attack, less disease infestation, good germination of seed and less termite attack as reported by 64, 18, 22 and 25 per cent of IRM farmers, respectively (Table 5.43). But a less number of IRM respondents (67) reported that the relative advantages of treated seed were observable. Out of these 72 per cent reported that there is apparent difference in pest attack, 19 per cent reported that t reated seed is saved from termite attack and germination of seed is good(18 per cent).

The IRM farmers not going for seed treatment is due to the perceived complexity. The complexities reported by the IRM farmers were: difficult to treat the seed (60 per cent), laborious (43 per cent), skill intensive (50 per cent), no previous experience (43 per cent) and time consuming (7 per cent).

Table 5.43: Attributes of Seed Treatment/Treated Seed

Attribute	*District Wise Percentage of Farmers*			*Overall Percentage of Three districts*
	Bathinda	*Ferozepur*	*Mansa*	
Relative advantage				
(i) Less/delayed jassid attack	75	44	62	64
(ii) Less disease infestation	22	25	5	18
(iii) Good germination of the seed	17	25	29	22
(iv) Less termite attack	14	31	38	25
n	**36**	**16**	**21**	**73**
Observability				
(i) Apparent difference in pest attack	80	62	67	72
(ii) Good germination	10	19	29	18
(iii) Less termite attack	10	19	33	19
n	**30**	**16**	**21**	**67**
Complexity of seed treatment				
(i) Difficult to treat the seed	33	57	100	60
(ii) Laborious	22	36	86	43
(iii) Skill intensive	55	36	71	50
(iv) No previous experience/ knowledge	55	21	43	43
(v) Seed treating chemicals not available locally	11	0	0	3
(vi) Time consuming	0	14	0	7
n	**9**	**14**	**7**	**30**

Multiple response. Decimals rounded up to nearest whole numbers.

5.7.4.2 Constraints in the Adoption of Seed Treatment

The major constraints reported by 93 IRM farmers, for not treating the seed were in terms of lack of knowledge (51 per cent), no previous experience (29 per cent), seed treating chemicals not available locally (5 per cent), chemicals being poisonous and laborious practice (2 per cent). The other reasons given by the IRM farmers were that there is no benefit of seed treatment (no observability) (18 per cent), IRM programme started late and farmers gained knowledge about seed treatment when sowing was complete (10 per cent). Table 5.44 gives an overview of the reasons and constraints for not treating seed.

5.7.5 Adoptability of ETL

The adoptability of ETL at farmers' level has been questioned by many as already discussed in this chapter. In the IRM villages, the scouts and trainers were imparting the skills for calculating ETL. The IRM farmers who had either adopted or observed trainers imparting training about ETL were asked to list the different attributes under relative advantages, observability, compatibility and complexity (Table 5.45).

Table 5.44: Constraints in the Adoption of Seed Treatment Practice (Other than treated seed)

Constraint/Reasons	*District Wise Percentage of Farmers*			*Overall Percentage of Three districts (n=93)*
	Bathinda (n=93)	*Ferozepur (n=34)*	*Mansa (n=29)*	
Constraints				
(i) Lack of knowledge	33	67	48	51
(ii) No previous experience	37	29	21	29
(iii) Non-availability of seed treating chemicals	17	0	0	5
(iv) Chemicals poisonous	7	0	0	2
(v) Laborious	0	0	7	2
Reasons other than constraints				
(i) No benefit of seed treatment	20	27	7	18
(ii) Gained knowledge about seed treatment during IRM-project	10	9	10	10

Multiple response. Decimals rounded up to nearest whole numbers.

Table 5.45: Attributes of Economic Threshold Level

Attribute	*District Wise Percentage of Farmers*			*Overall Percentage of Three districts*
	Bathinda	*Ferozepur*	*Mansa*	
Relative advantage				
(i) Reduced pesticides consumption/no. of pesticide applications	60	27	28	36
(ii) Reduced pesticide and spray expenditure	53	18	44	40
(iii) Need based/Rational use of pesticides	20	27	17	21
(iv) Effective pest control	13	9	11	11
(v) Increase in time interval between pesticide applications	3	23	33	17
(vi) Estimate of pest population	0	27	11	11
(vii) Regular field visits	3	0	0	1
(viii) Effectiveness of previous sprays estimated	0	5	0	1
(ix) For triai purpose	0	5	6	3
n	**30**	**22**	**18**	**70**

Contd...

Table 5.45–Contd...

Attribute	*District Wise Percentage of Farmers*			*Overall Percentage of Three districts*
	Bathinda	*Ferozepur*	*Mansa*	
Observability				
(i) Effective control	77	71	94	79
(ii) Saved on pesticide use/spray	40	33	18	32
(iii) Reduction on pesticide expenditure	27	0	0	12
(iv) Timely control of pest	0	5	0	1
n	**30**	**21**	**17**	**68**
I. Compatible with				
(i) Knowledge of ETL	73	50	44	56
(ii) Skill	44	22	30	32
(iii) Previous pest management experience	11	0	23	11
II. Not compatible with				
(i) Knowledge of ETL	9	7	19	11
(ii) Skill for calculating ETL	25	24	33	24
(iii) Previous pest management experience	47	78	77	67
(iv) Land holding	2	4	5	4
(v) Time	7	11	12	10
n	**45**	**46**	**43**	**134**
Complexity				
(i) Difficult to observe damage of the pests	4	9	26	13
(ii) Difficult to calculate	46	41	71	53
(iii) Labour intensive	34	38	77	53
(iv) Skill intensive	33	29	39	34
(v) Difficult to use in whole area	13	21	3	12
(vi) Time consuming	63	76	77	73
n	**24**	**34**	**31**	**89**

Multiple response.

The relative advantages of ETL were listed by 70 IRM farmers, who reported the relative advantages of ETL as: reduced pesticide consumption (36 per cent), reduction in pesticide expenditure (40 per cent), need based use of pesticides (21 per cent), increase in time interval between pesticide applications (17 per cent), estimation of pest population (11 per cent), effective pest control (11 per cent),trail purpose(3 per cent), effectiveness of previous spray estimated(1 per cent), and regular field monitoring (1 per cent).

The observability of these was reported by 68 IRM farmers and the observable effects were: saving on pesticide use (32 per cent) and effective control (79 per cent). But out of 134 IRM farmers, 67 per cent reported it to be non-compatible with: previous pest management experiences, 24 per cent with skill for calculating ETL, 11 per cent with knowledge of ETL, four percent with land holding and 10 percent with time. Out of 134 IRM farmers 56 per cent and 32 per cent reported it to be compatible with knowledge of ETL and skill, respectively. Only 11 per cent reported it to be compatible with previous pest management experience.

The other attributes other than compatibility which is important in case of ETL is the complexity. Out of 147 IRM farmers, 89 reported it to be complex and main complexity was difficult to calculate (53 per cent), labour intensive (53 per cent), skill intensive (34 per cent), difficult to observe (13 per cent), difficult to calculate (use) in whole area (12 per cent) and time consuming (73 per cent). Similar limitations of ETL have been reported by Zadox (1985), van de Fliert (1993) and others. In industrialized agriculture like the USA, the farmers employ scouts for calculating ETL, but all farmers are not able to pay the money for scouting. The requirements for the types and extent of population scouting to support decision making process needs to be determined and who will pay for it needs to be looked into (Norris *et al*, 2002).

5.7.5.1 Constraints in the Adoption of ETL

The constraints limiting the adoption of ETL were similar in terms of complexities. The constraints expressed by 88 IRM farmers, were that determining ETL is time consuming (26 per cent), lack of proper knowledge, comprehension and skill (20 per cent), laborious (22 per cent), pest population never being below ETL in case of ABW and whitefly (8 per cent).

One of the attributes of ETL expressed by farmers was 'risk' which was not forming the part of semi-structured questions related to attributes of ETL, but in case of open ended questions related to constrains 'risk' was reported as a constraint by 19 per cent of the IRM farmers (Table 5.46). The other reasons for not adopting ETL were that: insecticides are applied as preventive measure (7 per cent), their past experience and general observations were enough to take pesticide related decisions (33 per cent). No benefit of ETL was reported by two per cent of the 88 IRM farmers. The ETL not being adopted puts a question mark on the applicability of this practice at farmers' level in Punjab and taking it as indicator for determining the level of IPM adoption.

5.7.6 Adoptability of IRM Strategy for Insecticide Use

The reasons given by 117 IRM farmers, who had partially or completely adopted IRM strategy of insecticides use, are reported in Table 5.47. The cotton growers in Punjab have suffered heavily due to losses caused by insect pests mainly bollworms during the last 8 to 10 years, so the farmers were keen to know about the recommendations of the PAU for insecticide use. The reasons for partially or completely following IRM strategy was to reduce insecticide expenditure (34 per cent), reduce insecticide use (15 per cent), to rotate the insecticide during cropping season (10 per cent), to avoid tank mixing (11 per cent), and for trial purpose

(14 per cent). One of the important reasons reported by the IRM farmers was that IRM staff created credibility (12 per cent) and availability of the IRM staff for advice (6 per cent). Some of the farmers (5 per cent) reported that they used IRM strategy only for using correct dosage. The other reasons listed by the IRM farmers were: to avoid resistance in pests, to avoid resurgence of whitefly by avoiding synthetic pyrethroids, to increase time interval between two applications and to avoid first spray of monocrotophos 36SL to conserve natural enemies.

Table 5.46: Constraints/Reasons in the Adoption of Economic Threshold Level (Reasons other than constraints also included)

Constraint/Reasons	*District Wise Percentage of Farmers*			*Overall Percentage of Three districts*
	Bathinda (n=19)	*Ferozepur (n=36)*	*Mansa (n=33)*	*(n=88)*
(i) Time consuming	16	31	27	26
(ii) General observation enough to take decision	42	25	36	33
(iii) Lack of knowledge/ comprehension and skill	21	17	24	20
(iv) Risk involved	16	19	21	19
(v) Laborious and difficult to calculate frequently	21	17	27	22
(vi) Application of insecticides as a preventive major based on our past experience	5	11	3	7
(vii) Use pheromone traps	5	0	0	1
(viii) Pest population never below ETL in case of ABW and whitefly	0	14	6	8
(ix) Application of insecticides at egg stage of the pest	11	3	0	3
(x) No benefit of ETL	0	5	0	2

Multiple response. Decimals rounded up to nearest whole numbers.

5.7.6.1 Attributes of IRM Strategy for Insecticide Use

The relative advantage of the IRM strategy for insecticide use were reported by 113 IRM farmers, out of them 73 per cent and 85 per cent reported that less pesticide consumption and reduced pesticide expenditure, respectively, were the relative advantages (Table 5.48). The observability of these relative advantages were reported by 106 IRM farmers, who reported (84 per cent) that effective control of insect pest was observed by following the IRM strategy. Five per cent of the IRM farmers reported that when American bollworm and spotted bollworm were present at the same time, IRM strategy is ineffective and four per cent reported that by following IRM strategy no effective control of tobacco caterpillar was achieved (Table 5.48).

Table 5.47: Reasons for Adopting Insecticides Partially/Completely as per IRM Strategy

Reasons	District Wise Percentage of Farmers			Overall Percentage of Three districts (n=117)
	Bathinda (n=46)	Ferozepur (n=34)	Mansa (n=37)	
(i) To reduce insecticides expenditure	43	23	32	34
(ii) To reduce insecticides use	9	21	16	15
(iii) To avoid resistance in insect pests	4	6	0	3
(iv) To get effective and better results	9	12	0	8
(v) To rotate the insecticides	2	26	5	10
(vi) To avoid tank mixing of insecticides	9	15	11	11
(vii) Credibility created by IRM staff (selfless service)	13	12	11	12
(viii) Trial/experimentation of IRM strategy to observe its results	9	21	14	14
(ix) No previous experience of tank mixing of insecticides	2	0	0	1
(x) Credibility of PAU	7	0	0	3
(xi) To avoid first spray of monocrotophos 36 SL and replaced it with endosulfan 35 EC to conserve natural enemies	4	9	5	6
(xii) Availability of IRM staff for advice	4	9	5	6
(xiii) To use recommended insecticides and dosages	0	9	5	5
(xiv) To increase time interval between two insecticide applications	0	0	8	3
(xv) Judicious use of insecticides for good health	0	9	0	3
(xvi) To avoid spraying up to 60-90 days after sowing	2	3	0	2
(xvii) When *Helicoverpa* sp infestation less followed IRM strategy	2	6	5	4
(xviii) Followed IRM strategy when single pest present at a given time	2	0	2	2
(xix) Avoided synthetic pyrethroids to avoid resurgence of whitefly	2	6	3	3

Multiple response. Decimals rounded up to nearest whole numbers.

Table 5.48: Attributes of Recommended Insecticide Management Practices Under IRM Programme

Attribute	*District Wise Percentage of Farmers*			*Overall Percentage of Three districts*
	Bathinda	*Ferozepur*	*Mansa*	
Relative Advantage				
(i) Less pesticides consumption	83	51	65	73
(ii) Reduction in pesticides expenditure	89	75	88	85
n	**47**	**32**	**34**	**113**
Observability				
(i) Effective control of insect pests	82	93	78	84
(ii) Higher cotton yield	18	14	6	13
(iii) No effective control of tobacco caterpillar	4	7	0	4
(iv) When ABW+SBW present at the same time IRM strategy not effective	2	7	6	5
n	**45**	**29**	**32**	**106**
Compatibility				
Compatible with past experience (Partially included)	69	81	77	75
Complexity				
Knowledge intensive, laborious and literature needed at every stage of decision making	2	36	35	24
n	48	49	49	146

Multiple response. Decimals rounded up to nearest whole numbers.

Compatibility was partial to full in case of 75 per cent of the IRM farmers, but 24 percent termed it complex, as following IRM strategy for insecticide use is knowledge intensive, laborious and literature has to be consulted at every step (24 per cent).

5.7.6.2 Constraints in the Adoption of IRM Strategy for Insecticide Use

As was in the case of ETL, 30 per cent of the 86 IRM farmers reported that following the IRM strategy for insecticides use, the most important limiting factor was 'risk'. The farmers reported tank mixing of insecticides gives better results than IRM recommendation (24 per cent), and especially when American bollworm and spotted bollworm are present at the same time (14 per cent). Ten per cent of farmers also reported that when infestation of bollworm (ABW) and tobacco caterpillar is severe, tank mixing is the only option and an equal number of farmers reported that the advice given by dealers/other farmers was followed for taking the pesticide use decision (Table 5.49). Two per cent of the 86 IRM farmers reported that pesticide companies even lay the trial of using tank mixtures so experimentation and trial should be laid by the IRM staff to make them observe the results of the IRM strategy of insecticide use. The detail list of reasons and constraints are given in Table 5.49.

Table 5.49: Constraints in the Adoption/Partial Adoption of Insecticides as per IRM Strategy

Constraint/Reason	*District Wise Percentage of Farmers*			*Overall Percentage of Three districts*
	Bathinda (n=8)	*Ferozepur (n=40)*	*Mansa (n=38)*	*(n=86)*
(i) Risk involved in following IRM strategy	37	30	24	30
(ii) Decision of elders for tank mixing of insecticides	13	0	5	3
(iii) Followed other farmers/ dealers advice	0	5	18	10
(iv) Tank mixing of insecticides gives better results	25	28	21	24
(v) In case of two pests present at the same time, tank mixing needed (*eg.* ABW+SBW)	25	13	13	14
(vi) In case of severe infestation of ABW/TCP tank mixing of insecticides effective	13	13	8	10
(vii) Tank mixed initially before developing confidence in IRM staff	0	10	0	5
(viii) Endosulfan 35EC intoxicating	0	3	5	3
(ix) Time interval between 2 pesticides application decreased and spraying cost increases without tank mixing	0	15	5	9
(x) Pesticide companies lay trails of tank mixing/using mixed insecticides	0	3	3	2
(xi) Difficult to give up old habits	0	5	3	3
(xii) Trial should be laid by IRM staff for seeing is believing	0	3	3	2
(xiii) During full-moon night spraying pesticides essential	0	0	5	2

Multiple response. Decimals rounded up to the nearest whole numbers.

5.7.7 Adoptability Indices of Selected IPM Practices

Adoptability of selected IPM practices, namely timely sowing of cotton crops, cultivating recommended resistant varieties, seed treatment, ETL and following IRM strategy for insecticides use were measured (quantified) on a three point continuum scale as described in Chapter 4. IRM farmers were administered this scale to work out the adoptability indices. The attributes of a particular practice selected for the study are enlisted in Tables 5.50 and 5.51. However, while administering the scale, and collecting the data regarding constraints in the adoption, one important attributes 'risk' was mentioned by the IRM farmers in case of ETL and adopting IRM strategy for insecticide use. Risk and uncertainty as an attribute of an innovation affecting the adoption were mentioned by Becker (1970) and Zaltman *et al.* (1973). The non-inclusion of 'risk', as an attribute for calculating indices of ETL and IRM strategy of insecticide use might reduce the validity and reliability of the indices of these practices.

The attributes of selected cultural practices as measured on a three point continuum of "*complete, partial, and no*" are given in Table 5.50. The percentage of farmers who reported by way of cannot say/do not know have also be included in the table. In case of timely sowing, 91 per cent of the IRM farmers reported it having complete relative advantages and 90 percent reported that the benefits are completely observable, but 76 per cent of the farmers reported it to be partially compatible.

In case of cultivating PAU recommended resistant varieties, only 19 per cent of the IRM farmers reported these having complete relative advantage and 19 per cent reported these being completely compatible. Where as, a high percentage of 41 per cent, of the IRM farmers, reported that PAU recommended varieties do not have any relative advantage, and 48 per cent reported that these are not compatible (Table 5.50).

Seed treatment was far less compatible than treated seed. A very high percentage of IRM farmers reported that they were not aware about the compatibility and complexity of seed treatment(80,74 per cent, respectively), whereas in case of treated seed 81 per cent of the respondents reported it to be completely compatible. The relative advantages and observability were reported by a lesser percentage of the IRM farmers; where as 34 and 38 per cent of IRM farmers reported that seed treatment has no relative advantage and observability, respectively (Table 5.50).

The economic threshold level was reported to have complete relative advantage by 51 per cent of IRM farmers, but complete observability was reported by only 43 per cent of the IRM farmers, and equally high percentage of IRM farmers reported that they have no knowledge about its relative advantage (32 per cent) and observability (46 per cent). About 49 per cent of IRM farmers reported ETL being completely complex, and 12 per cent reported it to be partially complex (Table 5.51). More than half the percentage of IRM farmers (51 per cent) reported that ETL is not compatible with their farming system.

In case of following IRM strategy for insecticides use, fifty-five per cent of the IRM farmers, reported the strategy having complete relative advantage and 22 per cent reported it having partial relative advantage. The observability of benefits were complete (61 per cent) and compatibility was complete (45 per cent). But a high

Table 5.50: Attributes of Selected Cultural Practices Disseminated Under IRM Programme

Attributes	District Wise Percentage of Farmers												Overall Percentage of Three Districts (n=146)			
	Bathinda (n=48)				Ferozepur (n=49)				Mansa (n=49)							
	Complete	Partial	No	Can't Say	Complete	Partial	No	Can't Say	Complete	Partial	No	Can't Say	Complete	Partial	No	Can't Say
Timely sowing																
(i) Relative advantage	84	10	6	0	94	3	0	0	94	4	2	0	91	6	3	0
(ii) Compatibility	10	82	8	0	27	69	4	0	22	78	0	0	20	76	4	0
(iii) Observability	86	10	4	0	94	6	0	0	90	6	4	0	90	7	3	0
Recommended resistant varieties																
(i) Relative advantage	13	44	31	12	16	14	43	27	27	16	49	8	19	25	41	15
(ii) Compatibility	13	44	31	12	27	25	25	26	18	37	35	10	19	35	30	16
(iii) Observability	29	2	61	8	25	7	35	33	39	10	47	4	31	7	48	14
Treated seed/seed treatment																
(i) Relative advantage	65	10	21	4	12	20	37	31	31	12	43	14	36	14	34	16
(ii) Observability	63	0	31	6	16	16	37	31	37	6	45	12	39	7	38	16
Treated seed																
(i) Compatibility	90	0	0	10	59	14	12	15	94	0	2	4	81	5	5	9
Seed treatment																
(i) Compatibility	8	0	17	75	6	4	22	66	0	0	4	96	5	1	14	80
(ii) Complexity	19	0	4	77	18	10	12	60	10	2	2	86	16	4	6	74

Decimals rounded up to nearest whole number.

Table 5.51: Attributes of Economic Threshold Level and Insecticides Management

Attributes	District Wise Percentage of Farmers												Overall Percentage of Three Districts (n=146)			
	Bathinda (n=48)				Ferozepur (n=49)				Mansa (n=49)							
	Complete	Partial	No	Can't Say	Complete	Partial	No	Can't Say	Complete	Partial	No	Can't Say	Complete	Partial	No	Can't Say
Economic threshold level																
(i) Relative advantage	63	10	8	19	49	2	6	43	41	8	16	35	51	7	10	32
(ii) Observability	54	8	7	31	41	2	10	47	35	0	8	57	43	3	8	46
(iii) Compatibility	31	12	50	6	8	39	47	6	14	17	57	12	18	23	51	8
(iv) Complexity	40	10	42	8	55	14	23	8	53	10	23	14	49	12	29	10
Insecticide management																
(i) Relative advantage	86	12	2	0	31	35	20	16	49	20	22	8	55	22	15	8
(ii) Observability	90	2	2	6	45	14	23	18	49	16	12	23	61	11	12	16
(iii) Compatibility	83	15	2	0	26	35	35	2	26	35	37	2	45	29	24	2
(iv) Complexity	10	21	69	0	10	8	78	4	8	78	14	0	10	14	75	1

Decimals rounded up to nearest whole numbers.

percentage of the IRM farmers (75 per cent) reported that it is not complex (Table 5.50).

On the basis of these response categories, adoptability indices of different practices were calculated and the responses by way of can not say or do not know were not included for calculating the indices. The attributes which were positively related to adoption process were summed and the sum of attributes which were negatively related to adoption of innovation was subtracted from it, and the sum total was divided by the number of attributes. The formulae for calculating indices are given in Chapter 4.

In district Bathinda, the adoptability index of timely sowing worked out to be 0.81 (81 per cent), cultivation of recommended resistant varieties 0.42 (42 per cent), treated seed 0.80 (80 per cent), seed treatment 0.22(22 per cent), ETL 0.40 (40 per cent), and insecticide resistance management 0.65(65 per cent) (Table5.52). In district Ferozepur, the adoptability indices of above referred practices in the same sequence worked out to be 0.85(85 per cent), 0.42(42 per cent), 0.64(64 per cent), 0.15(15 per cent), 0.32(32 per cent) and 0.38(38 per cent). In case of district Mansa the adoptability indices were 0.83(83 per cent), 0.33(33 per cent), 0.63(63 per cent), 0.03(3 per cent) 0.27 (27 per cent) and 0.43(43 per cent) in case of timely sowing, cultivation of recommended resistant varieties, treated seed, seed treatment, ETL and insecticides resistance management, respectively (Table 5.52).

Overall in all the IRM villages of the three districts, the adoptability index of timely sowing worked out 0.83(83 per cent), cultivating recommended resistant varieties 0.39(39 per cent), treated seed 0.69(69 per cent), seed treatment 0.14 (14 per cent), ETL 0.35 (35 per cent) and insecticide resistance management 0.49 (49 per cent). The adoptability index of cultivating recommended varieties (other than *Bt* cotton) to decrease with the formal introduction of *Bt* cotton in Punjab in 2005 cotton growing season.

Against these adoptability indices, the actual adoption during 2004 was: timely sowing adoptability index 83 per cent actual adoption was 74 per cent of the farmers (this includes also the farmers who had partially sown in April); PAU recommended resistant varieties: adoptability index 39 per cent–actual adoption was 29 per cent farmers; seed treatment: adoptability index 14 per cent and actual adoption was six per cent, treated seed: adoptability index 69 per cent actual adoption was 72 per cent, ETL: adoptability index 35 per cent actual adoption was 44 per cent, IRM strategy for insecticide use: adoptability index 49 per cent–actual adoption was 42 per cent (partially following IRM strategy excluded) (Figure 5.28). The adoptability indices can give a rough estimate for predicting the adoption of a particular practice. The scale can be improved upon, and with more precise selection of attributes it can be applied to predict the adoption of different practices under IPM programmes. Differential weightage scheme based on the relative importance of an attribute for adoption of a particular practice will provide more precise estimates. But use of open ended questions along with the scale is very important to make farmers express their own opinion. As the attributes of 'risk' involved was reported by farmers in response to open ended questions.

Table 5.52: Adoptability Indices of Selected Cotton IPM Practices Disseminated Under IRM Programme

Practice	*Indices of Attributes*				*Adoptability Index*
	Relative Advantage	*Observ-ability*	*Comp-atibility*	*Complexity*	
Bathinda					
➢ Timely sowing	0.91	0.93	0.58	–	0.81
➢ Cultivation of recommended resistant varieties	0.22	0.66	0.39	–	0.42##
➢ Treated seed/Seed treatment	0.73#	0.67#	1.00 0.33	– 0.82	0.80 0.22
➢ Economic threshold level	0.83	0.85	0.40	0.49	0.40
➢ Insecticide resistance management	0.93	0.96	0.91	0.21	0.65
Ferozepur					
➢ Timely sowing	0.97	0.97	0.61	–	0.85
➢ Cultivation of recommended resistant varieties	0.32	0.42	0.51	–	0.42##
➢ Treated seed/Seed treatment	0.59#	0.35#	0.97 0.25	0.00 0.58	0.64 0.15
➢ Economic threshold level	0.87	0.79	0.29	0.68	0.32
➢ Insecticide resistance management	0.56	0.64	0.46	0.15	0.38
Mansa					
➢ Timely sowing	0.96	0.93	0.61	–	0.83
➢ Cultivation of recommended resistant varieties	0.35	0.24	0.41	–	0.33##
➢ Treated seed/Seed treatment	0.43#	0.47#	0.98 0.00	– 0.79	0.63 0.03
➢ Economic threshold level	0.69	0.81	0.26	0.69	0.27
➢ Insecticide resistance management	0.66	0.76	0.46	0.15	0.43
Overall					
➢ Timely sowing	0.95	0.94	0.60	–	0.83
➢ Cultivation of recommended resistant varieties	0.30	0.44	0.44	–	0.39##
➢ Treated seed/Seed treatment	0.58#	0.50#	0.98 0.19	– 0.73	0.69 0.14
➢ Economic threshold level	0.84	0.84	0.30	0.60	0.35
➢ Insecticide resistance management	0.72	0.79	0.61	0.17	0.49

The numbers of responses by way of can't say have not been taken into consideration for calculating indices.

#: It includes both "treated seed" and "seed treatment".

##: The adoptability index of resistant varieties most likely to decrease as a result of adoption of *Bt* cotton (refer Table 5.50).

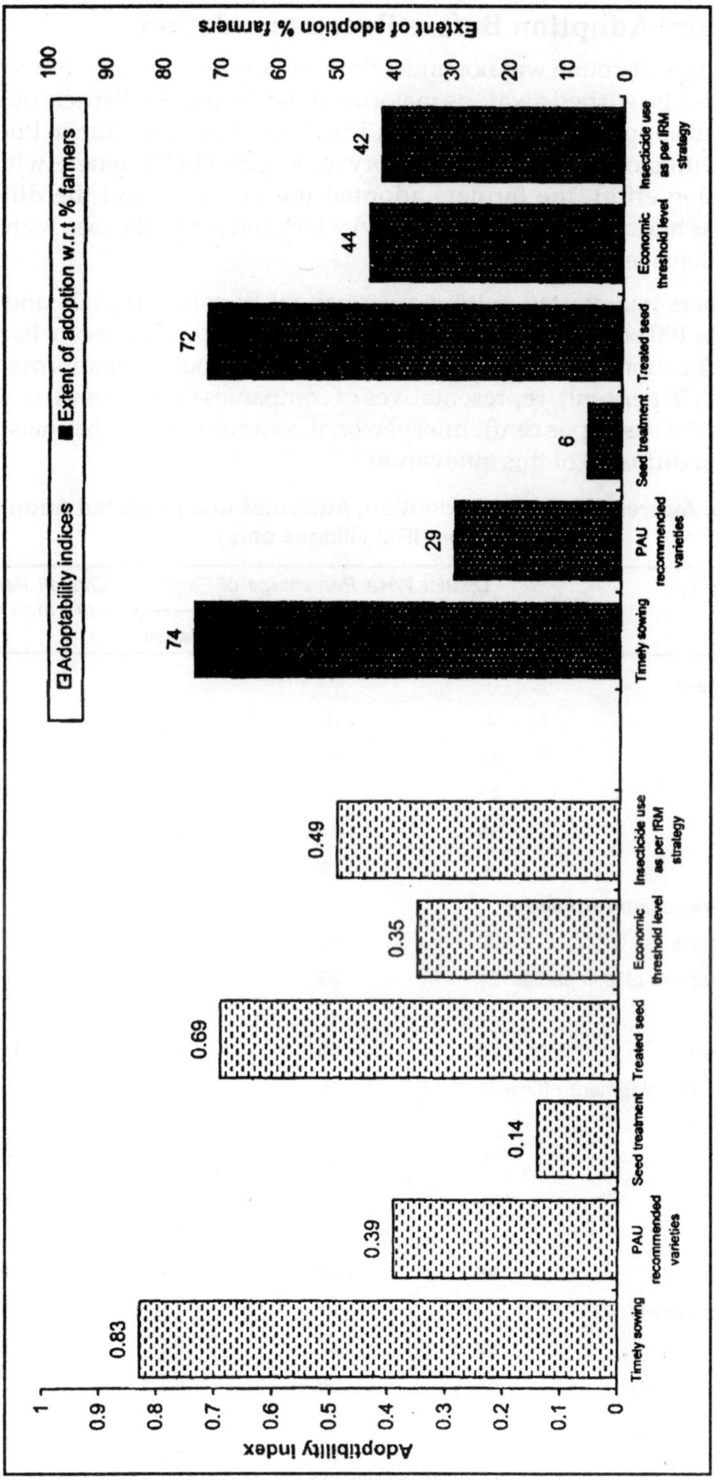

Figure 5.28: Adoptability and Adoption of Selected IPM Practices
a.i.: active ingredients

5.8 *Bt* cotton: Adoption Before Recommendation

Even though, Bt-cotton was not under the scope of present study, but at the same time it could not be wished away, as majority of the farmers in districts of Bathinda and Mansa had adopted Bt-cotton in 2004. The diffusion of Bt-cotton in Punjab is an example of 'Diffusion of Innovation Theory' of Rogers (1983), where without any public extension effort, the farmers adopted the Bt-cotton and the diffusion of innovation was horizontal (farmer to farmer) which can not be the case with complex knowledge intensive technologies like IPM.

The farmers had started getting aware about Bt-cotton in 2000 and by 2004 awareness was 100 per cent. The majority of the IRM farmers, 71 per cent had come to know about Bt-cotton in 2002 and 2003 (Table 5.53). The sources of information were other farmers (76 per cent), representatives of companies (19 per cent), newspapers (11 per cent), *Arthias* (5 per cent). Interpersonal communication channels were the main source of diffusion of this innovation.

Table 5.53: Awareness, Rate of Adoption, Attributes and Expected Adoption of *Bt*-Cotton (IRM villages only)

Category	*District Wise Percentage of Farmers*			*Overall Percentage of Three districts*
	Bathinda	*Ferozepur*	*Mansa*	
Year of awareness				
(i) 2000	4	0	0	1
(ii) 2001	15	0	10	8
(iii) 2002	42	20	27	30
(iv) 2003	35	41	47	41
(v) 2004	4	39	16	20
Source of awareness/information				
(i) Other farmers	77	73	78	76
(ii) Representatives of companies	21	20	16	19
(iii) *Arthia*	2	6	6	5
(iv) News papers	13	10	10	11
(v) Agriculture Development officer	2	0	0	1
Rate of adoption				
(i) 2002	8	0	4	4
(ii) 2003	33	0	14	16
(ii) 2004	98	38	80	72
n	**48**	**49**	**49**	**146**
Source of seed (2004)				
(i) Authentic	52	42	33	43
(ii) Not authentic	44	58	60	52
(iii) Can't say	4	0	7	5
n	**48**	**79**	**40**	**107**

Contd...

Table 5.53–Contd...

Category	*District Wise Percentage of Farmers*			*Overall Percentage of Three districts*
	Bathinda	*Ferozepur*	*Mansa*	
Attributes of Bt cotton#				
(i) Resistant to bollworms	96	100	98	97
(ii) Higher yield	83	87	93	87
(iii) Saving on pesticides expenditure	94	57	93	84
(iv) Easy to adopt	92	37	75	72
(v) Compatible	77	23	68	60
(vi) Not compatible	6	0	8	5
(vii) Timely wheat sowing	0	7	10	2
(viii) High cost of seed	8	10	15	11
(ix) More water requiring	46	17	18	29
(x) Higher fertilization dosage needed	8	0	0	3
(xi) Susceptible to CLCuV	0	0	3	1
(xii) Susceptible to tobacco caterpillar	6	3	0	3
(xiii) No benefit	4	3	8	5
n	**48**	**30**	**40**	**118**
Expected adoption of Bt cotton in 2005				
(i) With respect to area (in ha)	209.8 (67.02)	121.2 (28.30)	112.6 (51.73)	443.6 (46.26)
(ii) With respect to Percentage of farmers	100	78	80	89
n	**48**	**49**	**49**	**146**

#: Multiple response; CLCuV: Cotton leaf curl virus.

Figures in the parenthesis are the per cent area based on area under cotton during 2004.

Percentage of farmers rounded up to nearest whole number.

The rate of adoption of *Bt* cotton formed 'S' shaped curve (Figure 5.29) which is in agreement with Rogers (1983) diffusion theory. The Bt cotton being similar to green revolution technologies, so the adoption theory is applicable to such technologies. Against four percent adoption in 2002, rate of adoption multiplied four times in 2003 and in 2004 rate of adoption was 72 per cent. The attributes of *Bt* cotton as reported by IRM farmers were resistance to bollworms, higher yielding, saving on pesticide expenditure, timely wheat sowing (relative advantages and observability); easy to adopt and compatible (compatibility); high cost of seed, more water requiring, higher fertilizer dosages, susceptible to CLCuV and tobacco caterpillar (non compatibility),

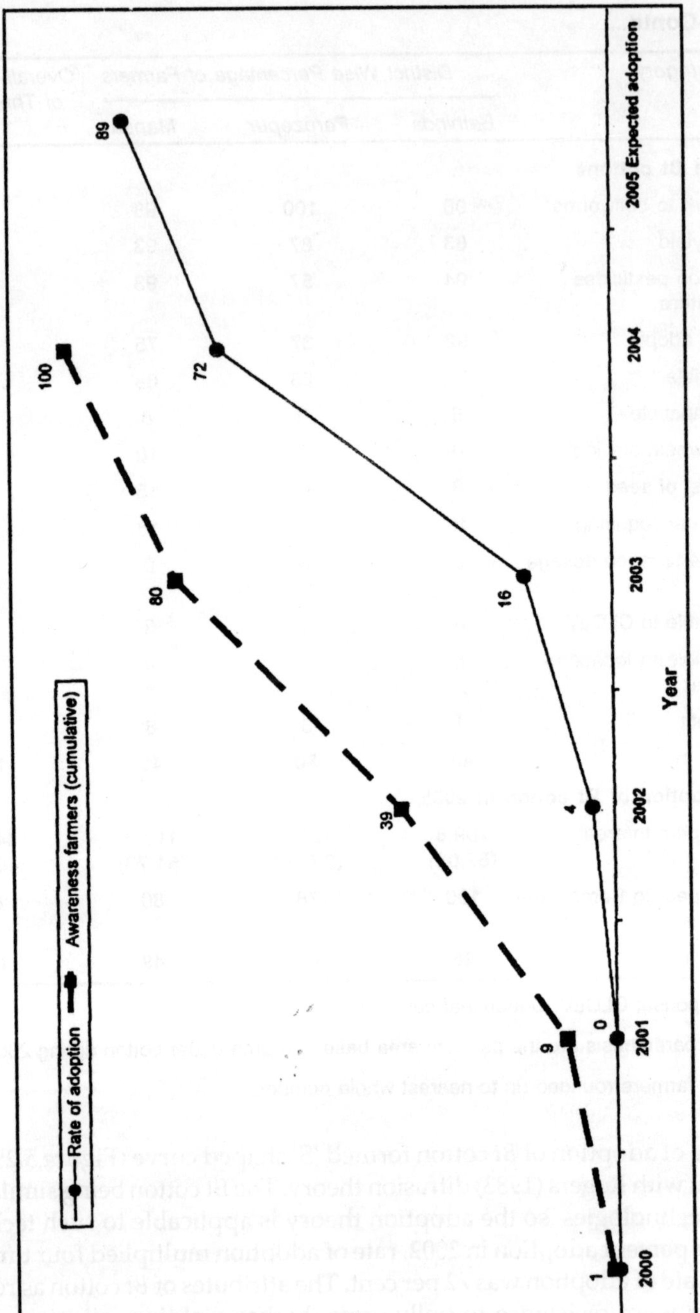

Figure 5.29: Awareness and Rate of Adoption of *Bt* Cotton in IRM Villages

but no complexity was reported by the farmers. The greater productivity of Bt-cotton was reported in the USA by $24.43 per acre and the increase in returns by $39.86 per acre (Boulter and Hilder 2002), but at the same time potential risks are also involved for example escape of insect resistance genes creating 'super weeds' maintaining rufugia to avoid breakdown of resistance (Boulter and Hilder 2002).

The majority of the IRM farmers (52 per cent) had not procured *Bt* cotton seed from authentic sources during 2004. Some farmers had even procured the seed from Gujarat state, and were sure of the authencity of seed (43 per cent).

On the basis of the information provided by IRM farmers, the likely percentage of adopters of Bt-cotton during 2005 might be around 89 per cent and the area under Bt-cotton is likely to increase to 46.26 per cent (Table 5.53) compared to 2004, which is an increase of 13 percent in terms of percentage of farmers and 19.56 per cent increase in terms of percentages of area (refer Table 5.14). In March 2005, six varieties of Bt cotton were recommended for cultivation in Punjab, these are: Ankur 2226, Ankur 651, RCH 134, RCH 317, MRC 6301 and MRC 6304.

5.9 Evaluation of the IRM Programme by Trainee Farmers

The training programmes in selected villages were attended from time to time to get an insight of the extension methodology followed for dissemination of the technology and the same was reported in section 5.3 of this chapter. Evaluation of the programme and training process by the trainee farmers would provide the user feed back to the stakeholders of IRM programme for its improvement.

5.9.1 Participation of Farmers in Training Programmes

The participation of the core group of ten farmers in each of the selected 15 villages was on regular basis, as 80 per cent of the IRM farmers regularly participated in group meetings (Table 5.54). Fifteen per cent of the IRM farmers were irregular and five percent rarely participated in group meetings. The reasons reported by the IRM farmers for not regularly attending group meetings were: lack of time (28 per cent), no fixed time schedule of the group meetings (31 per cent), very few group meetings conducted (24 per cent) and day time meetings (3 per cent). The infrequent group meetings being conducted was only in case of few villages.

5.9.2 Evaluation of the Training

The farmers rated the programme as good, as 82 per cent IRM farmers reported, that overall the programme was good, and 96 per cent reported that IRM programme was beneficial. Knowledge gain was reported to be the first reward of the programme. Knowledge gain in terms of insect pest identification was reported by 68 per cent of IRM farmers and knowledge gain about natural enemies was reported by 46 per cent of IRM farmers (Table 5.55). The knowledge gain about pesticide application was reported by 43 per cent, reduction/judicious use of pesticide 37 per cent, no tank mixing 19 per cent, no pesticide application until 60-70 days after sowing 21 per cent, regular field visit four per cent, ETL of pests seven per cent, and overall knowledge gain 16 percent of the IRM farmers (Table 5.55). The other benefits of the IRM programme as reported by the percentage of IRM farmers were reduced pesticide expenditure (32 per cent), timely advice (2 per cent), saved from pesticide dealers (2 per cent), and expert information at farmers' door steps (4 per cent).

Table 5.54: Participation of Farmers in Training Programme

Parameter	*Category*	*Percentage of IRM Farmers*
Participation in the group training	(i) Regular	80
	(ii) Not regular	15
	(iii) Rarely	5
	n	**146**
Reasons for not attending regularly#	(i) Lack of time	28
	(ii) Initially reluctant	7
	(iii) Other works	31
	(iv) No fixed time schedule for meetings	31
	(v) Very few meetings conducted	24
	(vi) Day time meetings	3
	(vii) Carelessness	3
	n	**29**

#: Multiple response. Decimals have been rounded up to nearest whole number.

In Table 5.55 are also listed the short comings of the IRM programme as reported by the IRM farmers. The major shortcoming was that farmers were not satisfied with the identification of natural enemies as 54 per cent of the IRM farmers reported about it. Printed material in the form of manual was provided to 87 per cent of the trainees and 13 per cent reported that manual was not provided to them. The manual was found informative (91 per cent), understandable language (91 per cent), comprehensive (84 per cent) by the IRM farmers but three per cent reported that difficult terms like ETL were not explained in detail. The trainees were by and large satisfied with IRM trainers but in case of group meetings only 56 percent farmers were satisfied (Table 5.55).

Table 5.55: Evaluation of the IRM-training Programmes by the Trainee Farmers

Parameter	*Category*	*Percentage of IRM Farmers*
Overall rating of the training programme (n=146)	(i) Good	82
	(ii) Average	10
	(iii) Poor	8
Beneficial/not beneficial (n=146)	(i) Beneficial	96
	(ii) Not Beneficial	4
Benefits# (n=146)	**I. Knowledge gain about**	
	(i) Insect-pest identifications	68
	(ii) Natural enemies	46
	(iii) Pesticides/its application	43

Contd...

Table 5.55–Contd...

Parameter	Category	Percentage of IRM Farmers
	(iv) Reduction/judicious use of pesticides	37
	(v) No tank mixing of pesticides	19
	(vi) No pesticided spray before 60-70 days after sowing	21
	(vii) Visiting field regularly	4
	(viii) ETL of pests	7
	(ix) Seeds	4
	(x) Overall knowledge gain	16
	II. Other benefits	
	(i) Reduced pesticide expenditure	32
	(ii) Increased time interval between insecticide sprays	1
	(iii) Timely advice	2
	(iv) Saved from pesticide dealers advice	2
	(v) Skill acquired	3
	(vi) Expert information at our door steps	4
Short comings# (n=140)	(i) Identification of natural enemies not satisfactory	54
	(ii) Skill to calculate ETL not imparted	2
Printed material (n=146)	(i) Provided	87
	(ii) Not Provided	13
Usefulness of printed material# (n=127)	(i) Informative	91
	(ii) Language understandable	91
	(iii) Comprehensive	84
	(iv) Difficult terms *viz.* ETL not explained in detail	3
About trainers# (n=146)	**Satisfied with their:**	
	(i) Communication skills	90
	(ii) Punctuality	84
	(iii) Technical capability	92
	(iv) Behaviour towards farmers	61
	(v) Conducting of group meetings	56

#: Multiple response; ETL: Economic threshold level.

Decimals have been rounded up to nearest whole number.

5.9.3 Sources of Information About Pesticides

Farmers in IRM villages reported that main sources of information regarding pesticide use during 2004 were IRM trainers, IRM manual and other literature. The percentages of IRM farmers reporting about the different sources of information is given in Table 5.56. For 75 per cent of the IRM farmers, main source of information was IRM project staff and 70 per cent reported that they consulted IRM manual for information regarding insecticides. It was reported by eight per cent and nine per cent of the IRM farmers that they consulted commission agents and pesticide companies' representatives, respectively for pesticide information. The IRM farmers who participated in the IRM training programmes mostly were the decision makers regarding the pesticide use, as 87 per cent of the respondents, reported that they were themselves decision makers about the pesticide use but in case of 26 per cent of the respondents, it was other family members who took the pesticide related decisions. The farmers who had purchased pesticide on credit were in some cases influenced by the commission agents as 14 per cent of the trained farmers, reported that commission agents did influence their decision in selecting a particular pesticide. The identity cards given to IRM farmers under IRM programme had helped them to get good quality insecticides as reported by the IRM farmers. The detail about the decision makers in selection and use of pesticides is given in Table 5.56.

Table 5.56: Source of Information and Decision Making About Pesticide Selection and Use in IRM Villages

Source/Decision Maker	*District Wise Percentage of Farmers*			*Overall Percentage of Three districts (n=146)*
	Bathinda (n=48)	*Ferozepur (n=49)*	*Mansa (n=49)*	
Source				
(i) IRM project staff	73	84	69	75
(ii) IRM literature	88	69	53	70
(iii) Other PAU literature	0	4	2	2
(iv) Other farmers	6	22	24	18
(v) Pesticides companies	2	10	14	9
(vi) Commission agents	0	10	14	8
(vii) State Agriculture Department	2	0	0	1
(viii) Previous experience	2	2	4	3
(ix) CIPMC-Jalandhar	2	0	0	1
Decision making				
(i)) IRM farmer himself	94	88	80	87
(ii) Family members	23	22	31	26
(iii) Other Farmers	6	6	12	8
(iv) Commission Agent	0	27	18	14
(v) IRM staff	48	59	55	54
(vi) Field workers	0	2	0	1
(vii) Pesticides company staff	0	4	10	5

Multiple responses.

Decimals have been rounded up to nearest whole numbers.

5.9.4 Suggestions of Farmers for Improvement of IRM Programme

The suggestions were sought from IRM farmers for making IRM programme more useful for them and the suggestions as given by IRM farmers are listed in Table 5.57.

Table 5.57: Suggestions of IRM Farmers for Improvement in IRM Training Programme

Aspect	*Suggestion*	*Percentage of IRM Farmers*
Group meetings (n=81)	(i) Training should be from seed to seed	73
	(ii) Impart skill for ETL and other difficult aspects	5
	(iii) Regularity of training	5
	(iv) Involve more farmers	4
	(v) More training by experts	5
	(vi) IRM programme should continue for 2 years	22
	(vii) Farmers need to cooperate	2
Field training (n=74)	(i) More field intensive and practical training	20
	(ii) Training should be in farmers' field only	66
	(iii) Identification of natural enemies and pests in the fields	9
	(iv) No need of lectures	5
Experimentation (n=121)	(i) Trial of different IRM recommendations	84
	(ii) Practical for experiential learning	61
Literature (n=26)	(i) Elaboration of difficult words like ETL, Kangi buti, etc.	12
	(ii) Update IRM book with more details about Tobacco caterpillar and its ETL and other aspects	73
	(iii) Fertilizer management in cotton	8
	(iv) Trade names of pesticides	15
	(v) Give the management of a particular pest along side with the detail of pest	8
	(vi) Recommendation not fit for our area	8
Timing of group meetings (n=41)	(i) Fixed day and time	73
	(ii) The Timing of group meetings should be morning or evening	73
	(iii) Improve the punctuality	7
Other suggestions (n=13)	(i) Cover IPM Programme for all major crops	39
	(ii) Aggressive publicity of the programme to compete with pesticide companies	15
	(iii) Marketing information	31
	(iv) Regulation and check on pesticide dealers	15
	(v) Farmers' tours to successful IPM farmers fields	8
	(vi) Use radio/TV/news paper for publicity	15

Multiple response. Decimals rounded up to nearest whole number.

A total number of 81 IRM farmers suggested improvements in group meetings and the main suggestion given by 73 per cent of the respondents was that training programme should be from seed to seed and 22 per cent suggested that it should continue in a particular village for at least two years, because in the first year they begin to understand different IPM concepts. The different suggestions with regard to group meetings were: impart skill training for ETL and other difficult aspects, regularity of weekly training, more training by experts and farmers should co-operate with IRM staff. A total number of 74 IRM farmers suggested about the improvements in field oriented training. Out of these, 66 per cent of the farmers suggested that training should be conducted in a farmer's field only and not at the *Panchayat Ghars* or other places, and nine per cent of the trainees suggested that identification of pests and natural enemies should be in the field and not by showing specimen or coloured posters. A small percentage of five per cent of trainees, suggested that there is no need of lectures as were being delivered by trainers in group meetings and 20 per cent suggested that training should be more field intensive and practical.

The suggestions of 121 IRM farmers were regarding experimentation, as 84 per cent out of them, suggested that trial of different IRM recommendations should be laid in farmer's field and 61 per cent suggested that training should be experiential learning based.

The written material provided to IRM farmers during 2004 was useful to them but certain suggestions were given by 26 farmers for making it comprehensive. They suggested that words like ETL, *Kangi buti* should be explained in written material. The IRM farmers also suggested that trade names of insecticides, fertilizers management, and management of a particular pest should be given along side the biology and life cycle of the pest. In the IRM manual being up dated during 2005, the management of pest has been given along side the life cycle of the pest. A small percentage of farmers also reported that the recommendations given in manual are not fit for their area.

Forty-one (28 per cent) IRM farmers suggested that the time and day of group meetings should be fixed and it should be conducted in evening or morning and punctuality should be improved. The number of villages in a district to be covered have been increased from 10 in each district in 2002 and 2003 to 15 villages in 2004 and 40 villages in 2005, and it is not possible for the district level trainers to cover all the villages in a week and impart meaningful training. Even though two trainers have been stationed in each district during 2005, but the logistic support is must for two trainers to move independently. A trainer should be assigned at the most 10 villages, which the trainer could cover in a week.

The IRM farmers gave other suggestions for improving the impact of programme. These included covering of IRM programme for all major crops of the season, like rice, to save on man power and funds, aggressive publicity of the programme to compete with pesticide industry, regular check on pesticide dealers, marketing information, use of radio/television and newspaper for popularizing the concept of IPM.

5.10 Inferences

The extension methodology for dissemination of IPM technology varies. In some cases, it is top-down approach, where as in other cases, it is participatory. Under Punjab conditions it should be a mix of top-down approach to compete with pesticide industry who have a large network to propagate their products by way of incentives, demonstrations, campaigns, banners, posters, etc. and participatory for making farmers confident (empowered) to take informed pest management decisions based on regular agro-ecoanalysis of their crop. The public extension agencies involved with IPM programme should adopt similar methods as adopted by chemical industry for popularizing the IPM philosophy, the street plays organized by IRM can help in that, but other methods need to be adopted for popularizing IPM and competing with pesticide industry. The top-down approach should be combined with participatory approach as far as educating the farmers is concerned. The participatory approach based on experiential learning, where farmer's field will act as a laboratory for discovering the new phenomena by farmers.

The increase in knowledge was mostly mentioned by the IRM farmers as the first reward, before other rewards like saving on pesticide expenditure. This is in line with the experience with other IPM programmes (Rossing *et al.*, 1985, Zadoks 1989b, van de Fliert 1993). The economists have also included the knowledge as an input for calculating the economic returns of IPM programmes (Feder and Quizon 1999). The theory on economic growth conditions considers knowledge as a third factor of production, after capital and labour (Romer 1990, Anonymous 1992). van de Fliert (1993) considers knowledge as determinants to relative advantages, but high adoption among trained farmers but low diffusion to non-trained farmers is not consistent with Rogers (1983) theory of diffusion. The IRM programme has resulted in gain in knowledge but not increased self confidence as farmers reported certain IPM practices to be risky. But the question arises what constitutes adoption of IPM? Based on the IPM continuum depicted by Kogan and Bajwa (1999) (Figure 5.30), the level of adoption of IPM can be considered at different levels of integration with each successive level incorporating the components of the preceding level. On this basis the IPM practices adopted by the IRM farmers is between level I and II. Farmers had adopted selective pesticides, pest detection, ETL and cultural practices. The level III was reported not having been achieved in any IPM programme (Norris *et al.*, 2002). Under the Punjab conditions, the indicators of IPM adoption can be: timely sowing, cultivation of resistant/tolerant varieties, treated seed, regular field agro analysis for observing pest and natural enemies, ETL and use of pesticides according to good agricultural practices.

The scope of this study being formative evaluation. Therefore, it is important to report on the knowledge gained about the process of IRM programme which led to the desired and undesired outcomes which can be fed into future implementation of the programme. Eventhough, knowledge was reported as the first reward of the IRM programme by the farmers, but acquired knowledge has not fully translated into application as the farmers were apprehensive about certain IPM practices. The experimentation as well as demonstration of the IPM practices would make farmers confident (empowered) about the IPM. Less attention was being paid to different

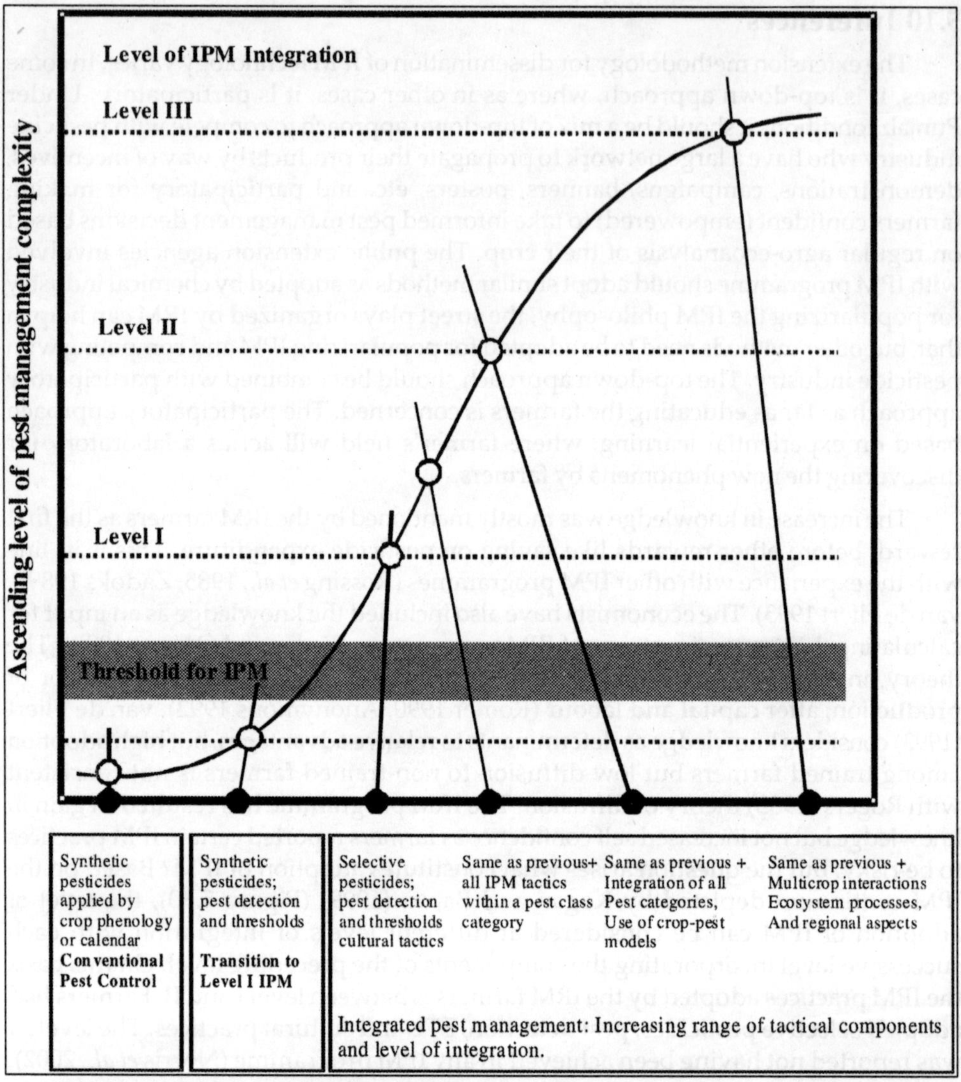

Figure 5.30: Diagram Depicting the IPM Continuum, and Showing the Relative Complexity of Different Levels of Integration
***Source*: Kogan and Bajwa, 1999**

cultural practices in the programme, as the programme lacked intensity until 70 days after sowing. The programme should give the same emphasis on the cultural practices as is being done in case of insecticide selection, dosage etc. Cultural and physical practices are considered important along with crop genetics in insect pest IPM programme than chemical control which is of minor importance (Norris *et al.*, 2002). Under Punjab conditions, cultural practices and use of insecticides based on IRM strategy are the major available components of IPM along with resistant varieties which includes *Bt* cotton.

Chapter 6

SUMMARY AND CONCLUSION

6.1 Introduction and Objectives

After independence, the Government of India during the period from 1947-1967, laid emphasis on enhancing input use in cotton crop to boost the productivity of cotton crop. The strategy was on enhancing high external inputs use like fertilizers and pesticides along with some varietal improvement efforts. These efforts led to a considerable increase in cotton area (7.8 mha) and production (5.3 m bales). Productivity however showed only a marginal improvement (114 kg ha^{-1}). The strenuous research and development efforts led to increase in cotton productivity to reach around 300kg ha^{-1} but still the productivity was much less than in other cotton growing countries of the world (around 600 kg ha^{-1}). The low productivity may be due to various factors, but the major factor is the cultivation of 65-70 per cent cotton crop in rain fed areas.

The productivity of cotton has significantly increased to 599 kg ha^{-1} since the introduction of *Bt* cotton in 2002. Presently out of around 9.0 million ha under cotton crop, around 6.9 million has is under *Bt* cotton. However, the productivity of cotton is still lower than the world average. One of the major factors for low productivity is the losses caused by insect pests, which increased from about 18 per cent under traditional agriculture in 1950s to around 50 per cent in 1990s despite increased use of pesticides in cotton crop. The pesticides use in cotton crop is about 50 per cent of the total pesticide consumption in India, even though area under cotton is five per cent of the total cropped area. It has come down to around 28 per cent in 2006–07 because of increase in the average of *Bt* cotton (Agranova, 2008). The pest scenario in cotton crop changed in north zone of India due to indiscriminate use of pesticides, which resulted in development of new pests, minor pests becoming major pests, pest outbreaks, pesticide resurgence and pesticide resistance.

As early as in 1975, the Indian Council of Agricultural Research sponsored a village level IPM project (Operational Research Project) to test and demonstrate the efficacy, practicability and economics of IPM in cotton. Government of India reoriented its plant protection strategy in 1985 and took a number of positive steps in promoting IPM. Different IPM projects and programmes were launched namely Regional Programmes for IPM in cotton by CAB International in 1993, FAO-European Union IPM Programme for cotton in 2000, Insecticide Resistance Management (IRM) Project in 2002 and recently in 2005 Sir Rattan Tata Foundation Project for cotton IPM was launched in Punjab. Despite all these efforts the percentage of insecticide consumption to total cost of cultivation kept on increasing.

Punjab has been the leader of the green revolution especially is case of increasing the productivity of rice and wheat, but the efforts have not resulted in achieving such a success in cotton crop. The farmers in cotton belt of Punjab used excessive insecticide in cotton that in some cases the insecticide costs were to the tune of 50 per cent of the total cost of cultivation. The consumption of pesticides in Punjab is the highest (7400 metric tones) in comparison to rest of India and more than half of it is used in cotton.

IRM is a component of IPM and this programme was launched in Punjab in 2002 under the broader IPM perspective to combat the development of resistance in pests through judicious use of insecticides, propagating adoption of primary cultural practices and cultivation of sucking pest and cotton leaf curl virus (CLCuV) resistant and short duration varieties. IPM is complex and the diffusion of it has been slow compared to conventional input technologies such as fertilizers and chemicals. To study the process of dissemination of IPM practices under IRM project, the formative process evaluation of the programme was planned, and the research problem entitled "Evaluation of Dissemination of Insecticide Resistance Management Technology in Cotton Crop in Punjab" was taken up with the following objectives:

1. To evaluate the farmers' training under IRM project in terms of training strategy and knowledge gain of the farmers.
2. To find out the adoption, adoptability and economic benefits of IPM practices in cotton crop disseminated under the IRM project.
3. To find out the constraints in the adoption of IPM practices disseminated under IRM project.

6.2 Materials and Methods

The quasi-experimental research design was adopted for conducting the formative evaluation of the IRM Programme. The study was conducted in all the three districts of Punjab where IRM programme has been implemented since 2002. These districts are Bathinda, Ferozepur and Mansa. Under this programme a total number of 45 villages were adopted (15 in each district), and out of which 15 villages were selected randomly-five from each district. Out of the 9 control villages-3 in each district, six villages were randomly selected (2 from each district). From each of the 15 IRM villages selected for the study, ten farmers forming the core group of trainees were purposively selected from each village. Thus the total number of IRM farmers selected was 150. In six non-IRM villages (Control villages) taken up for the study, ten farmers were selected randomly from the list of cotton growers provided by the IRM trainers. Thus the total sample size was 210 farmers (150 from IRM villages and 60 from non-IRM villages).

The IPM practices for cotton crop disseminated under IRM strategy were:

6.2.1 Cultural Practices

Cultural practices recommended under the IRM programme were: growing recommended resistant early maturing varieties, destroying alternate hosts of insect pest *viz. Kanghi buti, Peeli buti, Itsit*, recommended fertilizer dosages, use of farm yard manure, seed treatment, timely sowing, seed rate, plant spacing, and judicious use of irrigation water.

6.2.2 Chemical Practices Based on the IRM Strategy

These included: no insecticides application between 0-60 days after sowing, use of endosulfan for sucking pest management between 60-90 days after sowing, and use of synthetic pyrethroids for bollworms (early instars) in case of emergency, otherwise to avoid organophosphates/carbamates/pyrethroids until 90 days after sowing. Avoid endosulfan after 90 days after sowing and use synthetic pyrethroids/ organophosphates/carbamates for bollworm management between 90-110 days after sowing. Between 110-140 days after sowing, use profenophos/quinalphos/ triazophos for young larvae, chlorphyriphos/acephate for grown up larvae of American bollworm (ABW) or spinosad/indoxacarb if above insecticides fail to control grown up larvae of ABW. Use ethion/triazophos for control of whitefly. After 140 days after sowing, use chlorphyriphos/indoxacarb/spinosad/quinalphos as emergency operation against ABW. Use ethion or triazophos for control of whitefly.

The pesticide application based on economic threshold level (ETL) ETL for jassid: yellowing and curling along leaf margins on 50 per cent plants, ABW and spotted bollworm: 5 per cent damage in shed fruiting bodies, whitefly: 6 adults/leaf or appearance on honey dew on 50 per cent plants.

6.2.3 Conservation of Natural Enemies

Conserve natural enemies by not using any insecticide until 60 days after sowing and to avoid broad spectrum insecticides until 90 days after sowing.

Data were collected in three stages with the help of three research schedules. The pre- and the post-training, 29 item knowledge test was administered in the IRM villages. The base line data were collected at the time of administering pre-training knowledge test in the IRM villages, and separately in the non IRM villages at the begining of cotton growing season in 2004. The research schedule number two was in the form of structured questionnaire in a folder which was given to both the IRM and non-IRM farmers at the begining of cotton growing season in 2004 for recording the practices adopted and inputs used in cotton cultivation during 2004. These were collected back after the cotton growing season was over. The research schedule number three consisted of a three point continuum scale to measure adoptability, the open ended questions to record the attributes of the selected IPM practices, reasons for adoption of a particular practice and the constrains in the adoption of selected IPM practices. Research schedule number three was only administered in the IRM villages. The training programmes in the selected villages were attended from time to time from April 2004 to September 2004, and the observations were recorded in a diary. The data were collected from April 2004 to February 2005.

The data were analysed with the help of percentage, frequencies, mean, standard deviation, paired 't' test, two sample 't' test of unequal variance, Pearson's correlation coefficient by using 'SPSS' computer package.

6.3 Findings

6.3.1 Selection Procedure of Villages and Farmers Under the IRM Programme

The IRM project staff made the selection of villages based on the advice of local scouts, agricultural development officers of the state department of agriculture. The meetings of farmers were organized in short listed villages for the final selection of the villages. After consulting the village *Sarpanchs* and farmers' final selection of the villages was made. The villages nearby were selected as control villages. The farmers forming core group of ten trainees in each IRM village were selected on the basis of active participation of the farmers in the IRM programme.

6.3.2 Profile of the Respondents

There was almost no difference between IRM and non-IRM farmers with regard to their educational level. In Bathinda and Mansa districts the IRM farmers' average land holding was more than non-IRM farmers, but in Ferozepur district it was the opposite. There was no significant difference in case of average insecticides, applications, insecticides expenditure and yields between the IRM and non-IRM farmers in 2003 base line data.

6.3.3 Extension Methodology Followed for Dissemination of Technology

The extension methodology followed was a mix of transfer of technology (ToT) and participatory, limited to participation by consultation. The emphasis was also on countering the aggressive marketing of pesticides by the chemical industry and their agents, so much so that the IRM farmers developed confidence in the IRM staff and stopped seeking the advice of agents of pesticides companies. Street plays were organized in the villages for creating awareness about the misuse of insecticides. The weekly group meetings were conducted and printed literature and IRM manual was provided to the IRM farmers. Identification of insect pests and natural enemies was done in the farmers' fields. Majority of the IRM farmers regularly participated in the group meetings. More emphasis should be on experiential learning to make farmers confident in taking decisions rather than lectures. The information centres were set up in the IRM villages and the connectivity of farmers with the IRM staff was through telephones and local scouts deployed at each village. As consumption of insecticides in cotton is very high compared to rice and the chemical industry has a vast network so the combination of participatory and aggressive promotion of IPM strategy will work under the Punjab conditions.

6.3.4 Knowledge Gain of the Farmers About IPM

1. The IRM farmers' knowledge gain about different IPM practices was significant. There was significant difference in the knowledge of IRM farmers pre- and post-training in case of the three districts. The farmers of districts Bathinda had significantly higher knowledge than IRM farmers of Ferozepur and Mansa districts.
2. Even through significant correlation was observed between formal education and knowledge scores of the IRM farmers, but the gain in knowledge was independent of the formal education level of the farmers. Both low and high educated farmers had gained equally in knowledge about different aspects of IPM. Formal education of farmers was not a limiting factors in acquiring knowledge.
3. There was an overall gain of knowledge about different practices, except fertilizer dosages. The farmers gain in knowledge about resistant varieties, seed treatment, alternate hosts of spotted bollworm, irrigation management was not pronounced. More emphasis needs to be given to these aspects in future training programmes.
4. The percentage of the IRM farmers identifying the insect pests was high in pre test. They already possessed knowledge about jassid, whitefly and American bollworm. But pre-training knowledge about spotted bollworm, aphid and tobacco caterpillar was low. There was appreciable gain in the knowledge about spotted bollworm and tobacco caterpillar, as these insect pests were more prevalent in 2004.
5. Knowledge about natural enemies was low except in case of spiders, but post-training appreciable knowledge gain was observed except in case of predatory bugs. The population of these bugs over the years has decreased and very rarely these bugs were observed in cotton crop in 2003 and 2004.
6. The knowledge gain about different groups of insecticides was observed and knowledge gain about proper insecticides to conserve the natural enemies until 90 days after sowing was high. The knowledge gain about use of recommended insecticides, resurgence of whitefly, avoiding tank mixing of insecticides, rotation of insecticide molecules was appreciable.
7. The IRM Farmers knowledge about ETL was negligible as only one to three per cent of IRM farmers had pre-training knowledge about ETL of different insect pests. The knowledge gains about ETLs had taken place but the percentage of farmers possessing the knowledge of ETLs was less than the half.
8. Farmers awareness and knowledge about the insecticides resistance due to over use/misuse of pesticides increased after IRM programme, but a high percentage of IRM farmers attributed the reduction in efficacy of pesticides to poor quality.
9. The knowledge gain was reported to be the first reward of IRM programme by the IRM farmers.

6.3.5 Adoption of IPM Practices

1. There was not much difference in the adoption of different cultural practices between the IRM and non-IRM farmers with respect to fertilizer dosages, selection of varieties, seed treatment, irrigation, adoption of PAU recommended resistant varieties, seed rate and plant spacing. The IRM farmers' extent of adoption of chemical weed control was higher as compared the non-IRM farmers and correct level of adoption with respect to dosages was also higher. The low knowledge of the IRM farmers about these practices was the limiting factor in adopting recommended cultural practices. Under the Punjab conditions the level of integration of IPM is cultural control and threshold based insecticide application. In the absence of effective bio-agents, the emphasis needs to be given to crop manipulation techniques which reduce buildup of insect pests, resistant cultivars and thresholds. The stakeholders of IRM programmes should lay a greater emphasis to cognition and skills training of the farmers.
2. Farmers' not using recommended dosage of fertilizers was due to the fact that the knowledge about it was low and the farmers keep on experimenting on their own for finding the suitable fertilizer dosages for there area. The fertilizer management needs to be taken up in IRM programme in a purposeful way.
3. The extent of adoption of PAU recommended non-*Bt* resistant varieties was low. Over all, the PAU developed varieties with an exception of F-1378 were not adopted over a large area.
4. The insecticides application according to the IRM strategy (right insecticides, right dosage, against a particular pest) is as complex in cotton crop as following IPM. In the IRM villages farmers had not applied broad-spectrum insecticides in the beginning and the extent of use of endosulfan was more, which is relatively safer to natural enemies. But we need to think beyond endosulfan as there is a move to ban it. No insecticide applications were given until 60 days after sowing, which is one of the strategies of IRM for conserving natural enemies.
5. The IRM farmers had used significantly less tank mixtures than non-IRM farmers and had rotated the molecules of different insecticides as per the the IRM strategies. The overuse of synthetic pyrethroids was also less in the IRM villages.
6. The extent of adoption of ETL was low in the IRM villages, in non-IRM villages it was zero, inspite of the fact that ETL for jassid was recommended by PAU in 1979 and ETL of whitefly and bollworms was recommended in 1991. This proves that rate of diffusion of skill intensive and complex technology was zero. The extent of adoption of ETL in IRM villages being low signifies that skill intensive training needs to be imparted about sampling procedure in IPM programmes otherwise the adoptability is doubtful.

7. More needs to be done in IRM training programmes in the form of experimentation and demonstration to motivate farmers about importance of cultural practices in pest management and use of insecticides as per the IRM strategy.
8. The IRM farmers had applied significantly lesser insecticide application and used lesser insecticides per ha (technical grade material) in comparison to the non-IRM farmers. This has been the second reward of the IRM programmes.
9. The farmers having larger land holding had applied more number of insecticide applications. There was significant correlation between size of land holding and number of pesticide applications. There was no significant association between formal education and insecticide applications.

6.3.6 Economic Benefit of IRM Programme

1. The average input cost of cotton cultivation did not show any worthwhile difference expect for the insecticide expenditure, but the net income of IRM farmers was higher than non-IRM farmers. The IRM farmers, on an average had earned more net income per hectare.
2. The insecticide cost to percentage of total cost of cultivation was low in the IRM villages, but still the insecticide costs are higher if compared with previously documented figures.
3. In 2004 cotton growing season, the IRM farmers had saved more than Rs. 1000 ha^{-1} on insecticides as compared to the non-IRM farmers, and difference was significant. This is the economic reward of the IRM programmes, but the IRM farmers should be informed about the economic benefits of following IPM philosophy so that they develop a positive attitude about IPM.
4. There was no significant difference in the yield per unit area between the IRM and non-IRM farmers.

6.3.7 Adoptability and Constraints

1. The adoptability of timely sowing is very high but the limiting factor is shortage of canal irrigation at the time of sowing. The farmers were knowledgeable enough about the relative advantages of timely sowing and had observed the benefits of timely sowing with respect to reduced pest damage, higher yield, less expenditure, timely wheat sowing and higher wheat yield.
2. The adoptability of the PAU recommended resistant varieties is very low, as the number of varieties promoted by private sector are large in number for farmers to choose from who have strong economic interest, large budgets for research and development, aggressive sales promotion, advertising campaigns and provide free incentives. With the introduction of Bt cotton the adoptability of the PAU recommended varieties will decrease further.

3. The adoptability of seed treatment is low, farmers lacked comprehension about the benefits of seed treatment, as its advantages in terms of the benefits are not observable enough as in the case of pesticide application. The hybrid varieties are already treated and the farmers sowing hybrid varieties automatically sow treated seed. The training under IRM programme should emphasize in inculcating habit of seed treatment in case of non-hybrid varieties to delay jassid attack.
4. The adoptability of completely following IRM strategy for insecticides use is neither high nor low. The result demonstrations along with experimentation of IRM strategy of insecticides use will increase its adoptability. The constraint in following IRM strategy was that the farmers reported it to be risky-as risk attribute is associated with it, which negatively affects its adoptability. This strategy was also reported to be complex by the IRM farmers.
5. The adoptability of ETL is low, the adoptability of ETL at farmers' level have been questioned by biological as well as extension scientists. The constraints were its complexity and non-compatibility with farming practices. Complexity in term of skills to calculate. It was also termed risky by the farmers to wait for ETL before using insecticides.

6.3.8 *Bt* Cotton: Adoption Before Recommendation

1. The rate of adoption of Bt cotton was in vertical straight line, even before its recommendation for Punjab in 2005. The Bt cotton was being cultivated since 2002 in Punjab and the extent of area under Bt cotton was the highest in Bathinda, followed by Mansa and the Ferozepur. It was the most widely cultivated variety in 2004 except in district Ferozepur, where hybrid cultivation is low.
2. The channels of communications for diffusion were farmers (horizontal commutation). This again proves that simple technologies diffuse with or without public extension agencies, but complex knowledge intensive technologies like IPM do not follow Rogers model of diffusion.

6.3.9 Evaluation of IRM Programme by Trainee Farmers

1. The IRM programme was termed beneficial by IRM farmers and reported knowledge as the first reward of the training, before economic reward in terms of reduced pesticide expenditure. The knowledge has been considered as an input by economists for calculating impact of IPM programmes.
2. The drawbacks about the IRM programme as reported by IRM farmers related to more of lecturing by the trainers than imparting field based training.
3. The IRM printed literature in the form of manual was considered informative by the farmers but the detail about the difficult terms like ETL, etc should be explained.
4. IRM farmers suggested that the training should start intensively right from sowing operation and end with picking of cotton.

5. The timing of group meetings should be either in the morning or evening, but it was not possible for one district level trainer to cover 15 villages in a week, if morning/evening schedule is followed. One trainer should be given 8-10 villages in a district, which can be regularly covered in a week for meaningful and productive training session in the field than at central places in a village.
6. The IRM farmers suggested arranging result demonstration or experimentation of the IRM strategy of insecticides use for making them feel confident about the strategy.
7. It was suggested that IPM programmes should cover all the main crops to save on resources and man power, but a multiplicity of IPM programmes are being implemented by different agencies for a single crop serving the same clientele.

6.3.10 Conclusion and Recommendations

1. The sustainable practices like IPM are complex in nature, so the extension methodologies for promotion of IPM need of to incorporate a mix of facilitation of farmers with top-down support of agencies promoting IPM. Diffusion of technologies like IPM, with multiple components is not useful and applicable to IPM. The conjectures also invite further research that can complement the short coming of diffusion of innovation theory. The IPM interdependence on a set of practices, which are interrelated, needs a different approach for dissemination.
2. IPM system has interpretive flexibility–as to what constitutes the adoption of IPM at a given place and under given condition? What can be the indicators of the adoption of IPM? Under Punjab conditions, the indicator of IPM adoption is not only reduction in pesticide usage as farmers are using high number of pesticide application. In case of reduction in pesticide usage, it can be termed as rational pesticide use. The level of integration of IPM in Punjab can be adoption of primary culture practices, manual mechanical practices and sampling based use of insecticides in cotton, in the absence of availability of bio-agents.
3. The model of IRM in terms of deploying local scouts for maintaining link between scientist and farmers can be adopted in case of other IPM programmes.
4. The adoptability of certain IPM practices like ETL needs to be looked into by entomologists. The ETL of different insect pests recommended 25 years ago in case of jassid and 13 years ago in case of whitefly and bollworms should be again validated by the scientists. Due to change in insect pest scenario during the last two decades, are these ETLs still valid?
5. Promotion of sustainable practices like IPM is of great implication for extension system of the state. Are the persons associated with IPM well trained for its dissemination? The mass media support for popularizing IPM is basic requirement to create awareness, and technical support for all IPM programmes should come from the state agricultural universities (SAU).

6. The biological scientist should involve social scientists from the field of extension education and economics to generate relevant feed back which can be utilized in a cyclic way for improving the processes (inputs) to achieve the intended outcome.
7. There should be systematic evaluation of IPM programmes based on evaluation methodologies as discussed in Chapter 3. Both formative and summative evaluation is desired to conduct input, process, outcome and impact evaluation of the IPM programmes. The model suggested for IPM programme evaluation is given in Table 6.1. Input evaluation should be conducted with farmers participation on pilot basis to find out whether the technology is compatible, and what areas of IPM need more emphasis during the implementation phase. On the basis of input evaluation which will act a feedback, the IPM programme will be fine tuned. The outcome evaluation will provide the results worth of the programme in terms of the indicators listed in the model. Impact evaluation needs to be conducted after a time lapse of 2-3 years to find out the long term impact of IPM programmes.
8. The enforcement agencies in Punjab lacked the will to enforce the laws and regulate these in case of spurious insecticides, as reported by the farmers and as observed in case of unscripted seed of Bt-cotton being sold in Punjab in 2004, or even before that. This is a cause of concern.

6.3.11 Recommendations for Future Research

For Biological Scientists

1. The biological scientists from the field of entomology should conduct research for validation of economic thresholds of different insect pests of cotton under diverse condition to find out whether the ETLs recommended by the PAU in late 1970s and early 1990s are still applicable and valid due to change in pest scenario in Punjab.
2. Is economic threshold applicable when it is immeasurably low as in case of bollworms, and especially when bollworm complex is present at the same time? Some farmers reported that during bollworm complex infestation the damage in shed fruiting bodies is always higher than ETL, so multi-location trials should be conducted to work out ETL for bollworm complex.
3. IPM programmes must be developed to address the integrated management of all major pest categories (plant diseases, insect pests weeds), if the programmes are to be made more relevant to overall crop management. For this researchers (scientists) should not work in isolation, but in coordination. Integration of disciplines is to be adopted for developing and disseminating IPM technology.

For Extension Scientists

1. The IPM, with its implementation through various programmes provides an exciting area for research to extension education scientists. The two

Table 6.1: System Approach for Evaluation of IPM Programmes Based on Logic Model

	INPUTS →	ACTIVITIES OR PROCESSES →	OUTCOME →	IMPACT
METHOD OF EVALUATION	*INPUT EVALUATION*	*PROCESS EVALUATION*	*OUTCOME EVALUATION*	*IMPACT EVALUATION* Immediate and Long-term
TYPE OF EVALUATION	FORMATIVE	FORMATIVE	SUMMATIVE	SUMMATIVE
RESEARCH DEGIGN	Survey design Sample Explorative study for validation of IPM methods at farmers' level and extension methodology	Quasi-Experimental With/Without Before/During IPM	Quasi-Experimental and ex-post facto design With/Without, Before/After IPM	Ex-post facto Survey Design With/Without
TIME DIMENSION	T1 BEFORE IPM programme	T2 During the implementation of IPM programme	T3 At the end of the IPM programme	T4 After IPM programme
Evaluation indicators	Evaluation of IPM methods (content validation) 1. Whether it fits farmer by conducting pilot studies 2. Adoptability of IPM practices in terms of technological attributes 3. Adaptability of IPM practices 4. Implementation constraints Evaluation of implementation (extension)methodology budget and resources	Implementation of IPM programme 1. Experiential learning of farmers, eg. Agro-ecosystem analysis, sampling for ETL 2. Demonstration of results validated during input evaluation 3. Active participation of farmers and their reactions regarding IPM programme Whether IPM intervention yields desirable changes? Causal relationship	1. Gain in cognition and analytical skills 2. Change in attitude 3. Adoption of IPM practices which have been identified at input evaluation stage	1. Reduction in pesticide use 2. Reduction in pesticide expenditure 3. Increased population of natural enemies and reduction in population of insect pests 4. Higher input/output ratio 5. Multiplier effect(farmer to farmer diffusion)

types of research in the field of IPM are must, one–to evaluate the immediate and long term impact of all IPM programmes in Punjab. The villages covered under different IPM programmes right from Operational Research Project to the newly introduced IPM programmes, should be selected randomly, and second–to conduct three four evaluation of the current IPM programmes, a baseline study, process (formative evaluation), outcome and terminal impact evaluation.

2. The adoptability indices worked out need to be retested, and with more related attributes and differential weight-age scheme to find out whether these can be used to provide reliable results for predicting/estimating the adoption of an innovation.
3. The different project results as reported in different project reports need to be evaluated to find out whether the same level of IPM is being adopted by the farmers after the withdrawal of project intervention as reported in the project reports, and if not, what are the constraints/reasons for the same.

Appendices

Appendix I

Table V.1: Area Under Different Varieties of Cotton During 2004-05

Sl.No. Variety	Area Under Different Varieties (in ha)						Total of All the Districts	
	District Bathinda		District Ferozepur		District Mansa			
	IRM Villages	Non-IRM Villages	IRM Villages	Non-IRM Villages	IRM Villages	Non-IRM Villages	IRM Villages	Non-IRM Villages
I. American cotton								
Hybrids								
1. Ankur 651	6.8	4.2	7.2	0	17.2	8.6	31.2	12.8
2. Ankur 2226	0	0	1	0	27.6	2.4	28.6	2.4
3. Ankur 2534	0	0	0.8	0	5.6	0	6.4	0
4. Ankur 655	0	0	0	0	11.2	0	11.2	0
5. Ankur Sita	0	0	0	0	1.8	0	1.8	0
6. Ankur Jassi	0	0	0	0	2	0	2	0
7. RCH 134	13.1	0	0.8	0	2.4	2.4	15.58	2.4
8. RCH 317	4	0	1	0	0.4	0	5.4	0
9. RCH 138	0	0	0	0	2.8	0	2.8	0
10. White Gold	1.3	0	0	0	2	0	3.3	0
11. Dhaval-2	5.2	4.6	3.8	2	2	0	11	6.6
12. Birla 88	7.6	1.2	0	0	4	0.8	11.6	2
13. Nav Bharat 151	3.4	0	0	0	0	0	3.4	0
14. LHH 144	0.2	0	0	0	0	0	0.2	0
15. Sri Nath	0.5	0	0	0	0	0	0.5	0
16. Ammi	2.4	0	1.6	0	0.4	0	4.4	0

Contd...

Table V.1–Contd...

Sl.No. Variety	Area Under Different Varieties (in ha)						Total of All the Districts	
	District Bathinda		District Ferozepur		District Mansa			
	IRM Villages	Non-IRM Villages	IRM Villages	Non-IRM Villages	IRM Villages	Non-IRM Villages	IRM Villages	Non-IRM Villages
17. Vikram 11	11	3.6	0	0	0	0	11	3.6
18. Om 3	0.4	0	0	0	0	0	0.4	0
19. Krishi Rattan	2	0	0	0	0	0	2	0
20. Shanker	1.2	0	0	0	0	0	1.2	0
21. Sarpanch	1.2	0	0	0	0	0	1.2	0
22. Nath Rani	0	3.6	1	0	6.8	0	7.8	3.6
23. Ganga Kaveri	0	0	0	0	3.8	0.2	3.8	0.2
24. Anmol Rattan	0	0	5.2	0	0	0	5.2	0
25. Som Nath	0	0	0.8	0	0	0	0.8	0
26. Samrat	0	0	1.8	0	0.4	0	2.2	0
27. Roshni	0	0	1.2	0	0	0	1.2	0
28. Raj Rani	0	0	0.6	0	0	0	0.6	0
29. Anand	0	0	0.4	0	0	0	0.4	0
30. Amar Killa	0	0	0.4	0	0	0	0.4	0
31. Kohinoor	0	0	0.8	0	5.6	0.4	6.4	0.4
32. H 8	13.3	2	4	0	14	6.8	31.2	8.8
33. Rosa Kissan	0	0	0.8	0	0	0	0.8	0
34. Punjab Rattan	0	3.2	1.2	0	0	0	3.2	1.2
35. Nawab 72	0	0	0.2	0	0	0	0.2	0

Contd...

Table V.1–Contd...

Sl.No. Variety	Area Under Different Varieties (in ha)						Total of All the Districts	
	District Bathinda		District Ferozepur		District Mansa			
	IRM Villages	*Non-IRM Villages*	*IRM Villages*	*Non-IRM Villages*	*IRM Villages*	*Non-IRM Villages*	*IRM Villages*	*Non-IRM Villages*
36. AR 8108	0	0	0	2.4	0	0	0	2.4
37. Super Gold	0	0	0.8	0	0	0	0.8	0
38. Prateekh	0	0	0.8	0	0	0	0.8	0
39. Sri Ram	0	0	1.6	0	0	0	1.6	0
40. Super 2087	0	0	0.8	0	0	0	0.8	0
41. Narmada	0	0	0	0	0.2	0	0.2	0
42. Govinda	0	0	0	0	1.6	3.2	1.6	3.2
43. Bunny	2.4	0	0	0	0.8	1.6	3.2	1.6
44. JK 1	0	0	0	0	0.4	0	0.4	0
45. Star 505	0	0	0	0	1.4	1.2	1.4	1.2
46. Maruti	0	0	0	0	0.8	0	0.8	0
47. H 1001	0	0	0	0	0.6	0	0.6	0
48. Ajit 11	0	0	0	0	1.8	0	1.8	0
49. Karishma	0	0	0	0	2.4	0	2.4	0
50. Super Sada Bhar	0	0	0	0	4.6	2.2	4.6	2.2
51. Tulsi 37	0	0	0	0	0.4	0	0.4	0
52. Basant	0	0	0	0	0.4	0	0.4	0
53. Paras	3.4	4	0	0	1.2	0	4.6	4
54. Bolt	0	0	0	0	2.8	0	2.8	0

Contd...

Table V.1–Contd...

Sl.No. Variety	Area Under Different Varieties (in ha)						Total of All the Districts	
	District Bathinda		District Ferozepur		District Mansa			
	IRM Villages	Non-IRM Villages	IRM Villages	Non-IRM Villages	IRM Villages	Non-IRM Villages	IRM Villages	Non-IRM Villages
55. Sona Shakti	0	0	0	0	1	0	1	0
56. Jalwa	0	0	0	0	2.8	0	2.8	0
57. Tata 111	4	0	0	0	0	0	4	0
58. Neelam	0	0	0	0	0.8	0	0.8	0
59. Other hybrids	3.3	0.8	5.4	5.8	0.6	0.4	9.3	7
60. Gyan 5	0	1.6	0	0	0	0	0	1.6
II. Other Varieties								
1. F-1861	1.6	0	21.6	0	0	0	23.2	0
2. LH-1556	0.45	0	9.6	0	0.8	9	10.85	9
3. F-846	2	2.8	2.8	0	0	0	4.8	2.8
4. F-1378	7.05	2.2	34.4	27.2	9	5.4	50.45	34.8
5. Jhurar (F-414)	0	0.8	101.4	4	0	0	101.4	4
6. Rosianwali	4	0	36.4	16	0	0	40.4	16
7. Rawat Saria	0	0	62.4	18.4	0	0	62.4	18.4
8. Hissar -1117	0	0	0.6	0	0	0	0.6	0
9. Kikar Kheria	0	0	10	0	0	0	10	0
10. PK-54	2	0	10.8	9.2	0	0	12.8	9.2
11. F-1352	0	0	8.6	8.4	0	0	8.6	8.4
12. Raja Sikander	9.5	0.8	3.2	0	2	0.8	14.7	1.6

Contd...

Table V.1–Contd...

Sl.No. Variety	Area Under Different Varieties (in ha)						Total of All the Districts	
	District Bathinda		District Ferozepur		District Mansa			
	IRM Villages	Non-IRM Villages	IRM Villages	Non-IRM Villages	IRM Villages	Non-IRM Villages	IRM Villages	Non-IRM Villages
13. PK-104	0	0	1	0	0	0	1	0
14. Sikanderpura	0	0	2	0	6.4	0	8.4	0
15. Ganeshgaria	0	0	1.2	0	0	0	1.2	0
16. 1161	0	0	4.4	0	0	0	4.4	0
17. Ageta Sikender	1.6	2.8	0	0	0.6	1.4	2.2	4.2
18. Abni Sikender	0	0	0	0	2.4	0	0	2.4
19. Badshah	2	0	0	0.8	0	0	2	0.8
20. 2027	0	0	0	2	0	0	0	2
21. RG-8	0	0	0	1	0	0	0	1
22. Punjab gold	0	0	0	54.8	0	0	0	54.8
23. 2051	0	0	0	6	0	0	0	6
24. SML-8	0	0	35.8	14	0	0	35.8	14
III. Bt Cotton (undescript)								
1. Bt cotton	134.6	27.2	30	31.2	43.45	16	208.05	74.4
2. Bt cotton (1st progeny)	53.6	8.8	5.4	1.6	18	3.6	77	14
IV. *Desi* Cotton (LD 327, Moti & others)	7.9	1.6	2.6	3.6	0.4	0	10.9	5.2

Appendix II

Table V.2: Frequency of Different Cocktails (Mixtures) of Insecticides Used in Cotton Crop (Not recommended under IRM strategy or by the PAU) During 2004

Insecticides Mixture	*District Wise Frequency of Tank Mixtures*						*Total Frequency of Three Districts*	
	District Bathinda		*District Ferozepur*		*District Mansa*			
	IRM Villages (n=49)	*Non-IRM Villages (n=20)*	*IRM Villages (n=49)*	*Non-IRM Villages (n=20)*	*IRM Villages (n=49)*	*Non-IRM Villages (n=20)*	*IRM Villages (n=147)*	*Non-IRM Villages (n=60)*
Alpha-methrin 10 EC with:								
(i) Acephate 75 SP	0	0	7	12	0	0	7	12
(ii) Chlorpyriphos 20EC	2	0	0	0	2	0	4	0
(iii) Endosulfan 35EC	0	0	2	0	0	0	2	0
(iv) Ethion 50EC	11	18	57	51	28	23	96	90
(v) Monocrotophos 36SL	0	0	1	0	1	0	2	0
(vi) Profenophos 50EC	0	0	10	3	0	0	10	3
(vii) Quinalphos 25EC	0	0	4	0	0	0	4	0
(viii) Triazophos 40EC	13	11	88	48	80	38	181	97
Cypermethrin with:								
(i) Acephate 75SP	0	0	1	0	2	0	3	0
(ii) Chlorpyriphos 20EC	1	0	2	0	3	4	6	4
(iii) Endosulfan 35EC	0	0	1	0	0	0	1	0
(iv) Ethion 50EC	12	5	12	0	22	34	46	46
(v) Profenophos 50EC	0	0	4	0	0	0	4	0
(vi) Quinalphos 25EC	0	0	0	3	1	0	1	3
(vii) Triazophos 40EC	1	0	9	3	17	6	27	9
(viii) Monocrotophos 36SL	0	1	1	0	0	0	1	1

Contd...

Table V.2–Contd...

Insecticides Mixture	*District Wise Frequency of Tank Mixtures*						*Total Frequency of Three Districts*	
	District Bathinda		*District Ferozepur*		*District Mansa*			
	IRM Villages (n=49)	*Non-IRM Villages (n=20)*	*IRM Villages (n=49)*	*Non-IRM Villages (n=20)*	*IRM Villages (n=49)*	*Non-IRM Villages (n=20)*	*IRM Villages (n=147)*	*Non-IRM Villages (n=60)*
Delta-methrin 2.8EC with:								
(i) Endosulfan 35EC	0	0	1	0	0	0	1	0
(ii) Ethion 50EC	0	0	1	0	0	0	1	0
(iii) Triazophos 40EC	0	0	0	8	2	3	2	11
Fenvalerate 20EC with:								
(i) Acephate 75SP	0	0	21	10	1	0	22	10
(ii) Endosulfan 35EC	0	0	1	0	0	0	1	0
(iii) Ethion 50EC	1	0	6	4	0	0	7	4
(iv) Monocrotophos 36SCL	0	0	2	6	0	0	2	6
(v) Profenophos 25EC	0	0	1	0	0	0	1	0
(vi) Quinalphos 25EC	0	0	3	4	0	0	3	4
(vii) Triazophos 40EC	1	0	2	0	0	0	3	0
Lambda cyhalothrin +								
Ethion 50EC	2	0	0	0	0	0	2	0
Acephate 75SP +								
Profenophos 50EC	2	0	0	0	2	0	4	0
Ethion 50EC +								
Indoxacarb 15SC	1	5	0	0	0	0	6	0

Contd...

Table V.2–Contd...

Insecticides Mixture	*District Wise Frequency of Tank Mixtures*						*Total Frequency of Three Districts*	
	District Bathinda		*District Ferozepur*		*District Mansa*			
	IRM Villages (n=49)	*Non-IRM Villages (n=20)*	*IRM Villages (n=49)*	*Non-IRM Villages (n=20)*	*IRM Villages (n=49)*	*Non-IRM Villages (n=20)*	*IRM Villages (n=147)*	*Non-IRM Villages (n=60)*
Triazophos 40EC with:								
(i) Chlorpyriphos 36SL	0	2	0	0	1	0	3	0
(ii) Ethion 50EC	0	0	0	0	0	9	0	9
Monocrotophos + Ethion + Alphamethrin	0	0	0	0	0	4	4	0
Ethion + Triazophos + Alphamethrin	2	0	0	0	0	0	2	0
Endosulfan + Ethion + Alphamethrin	0	0	0	1	0	0	0	1
Triazophos + Ethion + Fenvalerate	0	0	0	1	0	0	0	1
Total frequency of tank mixtures of insecticides	49	35	244	154	162	121	455	311
Percentage insecticides applications as tank mixtures#	12.28	25.18	38.07	58.78	34.18	56.22	30.81	50.24

#: Based on total number of insecticide applications by all the respondent farmers.

BIBLIOGRAPHY

Adalla C B and Rola A C (1986) Participatory verification and technology generation of location specific IPM technology in Calamba Laguna: A pilot project. *Filipino Women in Rice Farming System*: 45-56.

Agnihotri N P (2000) Pesticide consumption in India-An update. *Pestic Res J* **12**: 1250-155.

Agarnova (2008) *Agrochemicals–Executive Review*, 19th Edition. Agarnova, U.K.

AICCIP (2008) All India Coordinated Cotton Improvement Project. *Annual Report* 2007–08, Project Coordinators Report.

Anonymous (1972) *Integrated Pest Management*. Council Environmental Quality, Washington, D.C. (Original not seen. Cited by Kogan M. 1998. *Annu Rev Entolmol* **43**: 243-70).

Anonymous (1987 and 1994) *Agricultural Statistics at a Glance*. Directorate of Economics and Statistics, Ministry of Agriculture, Government of India.

Anonymous (1991) *Handbook of Cost-benefit Analysis*. Department of Finance, Australian Government Publishing Service, Canberra, Australia.

Anonymous (1992) Economic growth: explaining the mystery. *The Economist* Jan **4**: 17-20.

Anonymous (1997) Guidelines for integrated production of arable crops in Europe. *IOBCwprs Bulletin Integrated Control in Oilseed Crops* **20**: 5-17.

Anonymous (1999) *Integrated Pest Management*. Food and Fertilizer Technology Center: An International Information Centre for Farmers in the Asia Pacific Region. **http://www/fftc.agnet.org.**

Anonymous (2001) *Agriculture Statistics at a Glancce-2001*. Ministry of Agriculture, New Delhi.

Anonymous (2003a) *Stastical Abstract of Punjab*. Economic Advisor to Government of Punjab.

Anonymous (2003b) (Original not seen. Cited by Russell D. 2004. Integrated pest management in less developed countries. In: Horowitz A R and Ishaaya I (eds) *Insect Pest Management: Field and Protection Crops.* pp 141-208. Springer, New Delhi).

Anonymous (2004) *Economic Survey 2003-04*. Ministry of Finance, Economic Division, Government of India, New Delhi.

Anonymous (2005) Agricultural centre for monitoring Indian economy. Economic Intelligence Service, Mumbai, India. **URL://www.cmie.com**

Anonymous (2006)Agriculture at a glance:2205-06. Information Service, Department of Agriculture, Punjab, Chandhigarh, India.

Atwal A S (1986) Future of pesticides in plant protection. In:*Proc Indian Natn Sci Acad* B 52:77-90.

Baca C P and Gonzalez C D (1994) Extension model and training method in IPM-maize transfer with small farmers in West Pacific Region, Nicaragua. *Proc Fifth International Congress of Integrated Pest Management Manejo – Integraode Plagas* **34**: 13-17.

Bambawale O M, Patil S B, Sharma O P and Tanwar R K (2004) Cotton IPM at cross roads. *Proc International Symposium on Strategies for Sustainable Cotton Production: A Global Vision*. pp 33-36.UAS, Dharwad, Karnatka.

Bannett C F (1977) *Analyzing Impacts of Extension Programmes*. Extension Service, USDA, Washington DC.

Barik A, Singh R P and Joshi S S (2002) *Technology Mission on Cotton in Nutshell*. Directorate of Cotton Development, Mumbai.

Becker (1970) (Original not seen. Cited by Zaltman G, Duncan R and Holbek J .1973. *Innovations and Organizations*. pp 32-40. John Wiley and Sons, New York).

Bell J (1983) *Contemporary Social Welfare*, Macmillan, New York. (Original not seen. Cited by Patton M Q. 1997. *Utilization Focused Evaluation*. Sage, Thousand Oaks).

Bentley J W and Andrews K L (1991) Pests, peasants and publications: Anthropological and entomological views of an integrated pest management programme for small scale-scale Honduran farmers. *Human Organisation* **50**: 113-124.

Bentley J W and Andrews K L (1996) Trough the road blocks: IPM and Central American small-holders. *Sustainable Agricultural Programme, Gatekeeper Series 56*. International Institute for Environment and Development, London.

Bentley J W, Rodriguez G and Gonzalez A (1993) Science and people: Honduran Campesinos and natural pest control interventions. In: Buckles D (ed) *Gorrasy Sombreros: Caminos hacia la Colaboracion Centre Tecnicos y Campesionosia*. Dept. of Crop Protection, El Zamarano, Honduras

Bernardo E N (1993) Adoption of IPM approach in crop production: A researcher's view. *Philippines Entomol* **9**: 175 -185.

Bhatia B S (1990) *Adoption of Farm Mechanization in a Developing Economy*. Daya Publishing House, Delhi.

Bhuiyan A K, Chau L A and Calilung V J (1992) The use of economic threshold levels: a sustainable strategy to pest management. *Ann of Bangladesh Agric* **2**: 25-30.

Birthal P S and Jha D (1997) Socioeconomic impact analysis of integrated pest management programmes. *Proc National Symposium on IPM in India: Constraints and Opportunities*, Oct 23-24, IARI, New Delhi.

Bleicher E, Silva A L, Calcagndo G, Nakano O and Freire E C (1979) Sistema de controle das pragas do algodociro paras a regiao centro sul do Brazil *EMBRAPA/ CNDPCRC Tee* 2. pp 21. (Original not seen. Cited by Ramalho F S. 1994. *Annu Rev Entomol* **39**:563-78).

Bloom BS (1979) (ed) *Taxonomy of Educational Objectives*. Longman Group Ltd, London.

Boulter D and Hilder V A (2002) Transgenic crops (annual). In: Pimental D (ed) *Encyclopedia of Pest Management*. pp 846-849. Marcel Dekker Inc, New York.

Braun A, vande Fliert E, Wheatley C, Prains G and Widodo Y (1995) Improving profits from sweet potato by linking IPM with diversification of markets. *CIP Circular* **21**: 8-15.

Brook K D and Hearn A B (1990) The SIRACTAC pest management computer program: Program content. *CSIRO Div Plant Industry Tech Paper*. pp 108.

Brunner I and Guzman A (1989) Participatory evaluation: A tool to assess projects and empower people. In: Hendricks M and Conner R F (eds) *International Innovation in Evaluation Methodology*. pp 9-18. New Direction for Programme Evaluation No 42. Jossey-bass, San Francisco, USA.

CABI (2000) *Learning to Ccut the Cchemicals in Cotton: Case Studies and Exercises in Farmer Focused Cotton IPM from Around the World*. Commonwealth Agricultural Bureau International-Bioscience, UK.

Campbell D and Erlebacher A (1970) How regression artifacts can mistakenly make compensatory education programme look harmful. In: Heltmith J (ed) *The Disadvantaged Child Vol 3 Compensatory Education: A National Debate*. Brunner, New York. (Original not seen. Cited by Kosloski K. 2000. *Encyclopedia of Sociology* 1. pp 861-869. Macmillan Refernces, New York USA).

Campbell D and Fiske D (1959) Convergent and discriminate validity by the multitrait-multimethod matrix. *Psychological Bulletin* **56**: 81-105.

Campbell D and Stanley J (1963) *Experimental and Quasi-Experimental Designs for Research*. Rand Mc Nally, Chicago, USA. (Original not seen. Cited by Kosloski K. 2000. *Encyclopedia of Sociology* 1. pp 861-869. Macmillan References, New York, USA).

Carson R (1962) *Silent Spring*. Fawcett Crest, New York.

Casley D J and Kumar K (1987) *Project Monitoring and Evaluation in Agriculture*. The John Hopkins University Press, Maryland.

Caswell M, Fuuglie K, Ingram C, Jans S and Kascak C (2000) *Adoption of Agricultural Production Practices- Lesson Learned from the US Department of Agriculture Area Studies Project*. Economic Research Services, USDA. **http://www.usda.org.**

Chandra M (1998) *Agriculture: Debt Drives Indian Farmers to Suicide*. **http://www.oneworld.net.**

Chin H, Othman Y, Loke W H and Rehman S A (1992) National Integrated Pest Management in Malaysia. *Integrated Pest Management in the Asia Pacific Region*: 191-209.

Chung D K, Pinucs J, Rola A and Widawsky D (1999) Assessment of household and village level impacts of IPM. In: Waibel H, Fleisher G, Kenmore PE and Feder G (eds) *Evaluation of IPM Programmes – Concepts and Methodologies*. pp 9-12. Pesticide Policy Project Publication Series 8, Hanover.

CICR (1994) *Annual Report*. Central Institute of Cotton Research, Nagpur, India.

CICR (1998) *Annual Report*. Central Institute of Cotton Research, Nagpur, India.

Cirio U (1992) Integrated pest management for olive growers in Italy. In Haskell PT (ed) *Research Collaboration in Europe IPM Systems*. pp 47-55. BCPC Monograph 52, BCPC Farnham.

Collins G S, Lacewell R D and Norman J (1979) Economic implications of alternative cotton production practices. Texas Lower Rio Grande Valley. *Southern J Agril Econ* **11**: 79-82.

Condra G P, Lindsely K E and Neeb CW (1975) Proposal for an upland cotton demostration in Reeves and Pecos counties. (Original not seen. Cited by Lacewell R D and Taylor C R. 1980. *Proc Seminar and workshop*. pp 284-302. CICP–USAID).

Cook T (1993) A Quasi-sampling theory of the generalization of causal relationship. In: Secherst L and Scott A (eds) *Understanding the Causes and Generalization about them: New Directions for Program Evaluation*. Jossey-Bass, San Francisco, USA

Cook T and Campbell D (1979) *Quasi Experimentation: Design and Analysis Issues for Field Settings*. Rand Mc Nally, Chicago, USA.

Coutt J (1997) Exposing the tiger. A fresh look at the evaluation of extension. *Proc 2nd Australasian Pacific Extension Conference*, Albury, New South Wales, Australia.

Cronbach L (1982) *Designing Evaluations of Educational and Social Programs*. Jossey-Bass, San Francisco, USA.

Cruz V R (1991) Algodao resultados do MIP, manejo integrado pragas no Estado do Sao Paulo em regio infestada pelo bicudo ano agricola 1990/91. *CATI Doc Tec* **89**:20.

Dent D (1995) *Integrated Pest Management*. Chapman and Hall, London.

Deshler D (1998) Evaluating extension programmes. In: Sawanson B E, Bentz R P and Sofranko A J (eds) *Improving Agricultural Extension: A Reference Manual*. Food and Agriculture Organization, Rome. **http://www.fao.org.**

Dhaliwal G S and Arora R (1993) Changing status of insect pests and their management strategies. In: Gill K S, Dhaliwal G S and Hansara B S (eds) *Changing Scenario of Indian Agriculture* pp 98-145. Commonwealth Publisher, New Delhi.

Dhaliwal G S and Arora R (1996) An estimate of yield losses due to insect pests in Indian agriculture. *Indian J Ecol* **23**: 70-73.

Dhaliwal G S and Arora R (2001) *Integrated Pest Management: Concepts and Approaches.* Kalyani Publishers, Ludhiana.

Dhaliwal G S and Arora R (2002) Estimation of losses due to insect pests in field crops. In: Sarath Babu B, Varaprasad K S, Anitha K, Prasada Rao R D V J, Chkrabarty S K and Chandurkar P S (eds) *Resource Management in Plant Protection* Vol 1. pp 12-23. Plant Protection Association of India, Hyderabad.

Dhaliwal G S, Arora R and Dhawan A K (2004) Crop losses due to insect pests in Indian agriculture: An update. *Indian J Ecol* **31**: 1-7.

Dhaliwal G S, Arora R and Heinrichs E A (1998) Insect pest management from traditional to sustainable approach. In Dhaliwal G S and Heinrichs E A (eds) *Critical Issues in Insect Pest Management*. pp 1-15 Commonwealth Publisher, New Delhi.

Dhaliwal G S, Arora R and Sandhu M S (1992) Operational Research Project at the door of cotton growers. *Farmer Parliament* **27**: 15-16.

Dhawan AK (1999) Major insect pests of cotton and their integrated management. In: Upadhay R K, Mukerji K G and Dubey OP (eds) *IPM Systems in Agriculture* Vol 6– *Cash Crops*. pp 165-225. Aditya Books Pvt Ltd, New Delhi.

Dhawan A K (2005) *Project Report: Dissemination of Insecticide Resistance Management*, 2004-05. Department of Entomology, Punjab Agricultural University, Ludhiana.

Dhawan A K, Sohi A S, Sharma M, Kumar T and Brar K S (2004) *Insecticide Resistance Management in Cotton*. Punjab Agricultural University, Ludhiana.

Directorate of Plant Protection, Quarantine and Storage (2001). (Original not seen. Cited by Dhaliwal G S, Arora R .2001. *Integrated Pest Management- Concepts and Approaches*. pp 270. Kalyani Publishers, Ludhiana).

Drost D, Long G, Wilson D, Miller B, Campbell W (1996) Barriers to adopting sustainable agricultural practices. *J Ext* **34**: 1-5.

Dudani A T and Sengupta S (1991) *Status of Banned and Bannable Pesticides*. Voluntary Health Association of India, New Delhi: 83.

Escalada M M and Heong K L (1994) New developments and need for training in IPM. *Proc 16th session of FAO/UNEP Panel of Exerts on Integrated Pest Control* 25-29 April, FAO, Rome.

Eveleens K G (2000) The FAO/EU cotton IPM programme in Asia: Problems and prospects. *Proc of the Brighton Crop Protection Conference 2000: Pest and Diseases.* pp 185-192. British Crop Protection Council, Farnham Surrey.

FAO (1968) *Report of the First Session of FAO Panel of Experts on Integrated Pest Control*, FAO: 19, Rome.

FAO (1996) *Country Briefs. Programme Advisory Committee Meeting*: FAO-ICP for IPM in Asia, February 06-09,1996, Hyderabad, India.

Farrington J and Nelson J (1997) *Using Longframes to Monitor and Review Farmer Participatory Research: Network paper* 73. Overseas Development Institute, London.

Feder G and Quizon J (1999) Integrated pest management (IPM) in Asia: Are there real returns to IPM and its diffusion? In: Waibel H, Fleisher G, Kenmore P E, Feder G (eds) *Evaluation of IPM Programs – Concepts and Methodologies*. pp 2-8. Pesticide Policy Project Publication Series 8, Hannover.

Fernando M H J P (1996) *Sri Lanka: National IPM Programme: A Country Brief.* Programme Advisory Committee Meeting, FAO-ICP for IPM in Asia, 06-09 February 1996, Hyderabad, India.

Figueredo A (1993) Critical multiplism, meta-analysis and generalization: An integrative commentary. In: Sechrest L (ed) *Programme Evaluation: A Pluralistic Enterprise*. Jossey-Bass, San Francisco, USA.

Fitt G P (1994) Cotton pest management: Part 3–An Australian perspective. *Annu Rev Entomol* **39**: 543-62.

Fliegel F C and Kivlin J E (1962) *Differences Among Improved Farm Practices as Related to Rate of Adoption.*Pennsylvania Agricultural Experiment Station, Research Bulletin 691.(Original not seen. Cited by Rogers E M. 1983. *Diffusion of Innovations.* The Free Press, London).

Fliegel F C *et al* (1967) *Innovation in India: The Success or Failure of Agricultural Development Programmes in 108 Indian Villages.* National Institute of Community Development, Research Report 9, Hyderabad.

Frisbie R E (1985) Consortium for Integrated Pest Management (CIPM) organization and administration. In: Frisbie R E and Adikisson P L (eds) *Integrated Pest Management on Major Agricultural Systems.* A & M Univ Texas.

Frisbie R E (1994) Integrated pest management. *Encyclopedia of Agricultural Science.* **2**: 569-578.

Frisbie R E and Adkisson P L (eds) (1985) *Integrated Pest Management on Major Agricultural Systems,* 743. Texas A & M Univ, Texas Agric Exp Stn, 743.

Gaines J C (1957) Cotton insects and their control in the United Sates. *Annu Rev Entomol* **2**: 319-38.

Galvan C and Kenmore P E (1991) IPM made easy. *World Edu Inc Reports Local Action for Global Change* **29**: 8-10.

Goldstein I L (1986) Evaluation procedures. In: Goldstein I L (ed) *Training in Organizations: Need Assessment Development and Evaluation.* Monterey: Brooks/ Cole, California.

Goodell G E (1984) Challenges to integrated pest management research and extension in the Third World: Do we really want IPM to work? *Bull Ent Soc Am* **30**: 18-26.

Green J C (1988) Stakeholder participation and utilization in programme evaluation. *Evaluation Rev* **12**: 91-116.

Groot A E and Boon C A M (1992), *Monitoring en Evaluatie: Een Werkplan Voor de Ontwikkeling van Monitoring en Evaluative in Gendergevoelige Procesmatige Ontwikkelings programma's*. Vakgroep Voorlichtings kunde in samen werkingmet SNV, Wageningen (Original not seen. Cited by van de Fliert E. 1993. *Integrated Pest Management: Farmer Field School Generate Sustainable Practices*. Wageningen Agricultural University Papers 93.3, Wageningen)

Haque M M, Sumaya B R, Martinez V A and Marzo M M (1988) The rural Filipino an extension aide (RUFEA): A pilot project in the extension of crop protection in rice. *Filipino Women in Rice Farming Systems*: 217-230.

Hasen M (1987) *Escape from Pesticide Treadmill: Alternatives to Pesticides in Developing Countries*. Institute for Consumer Policy Research, Consumers Union, New York.

Head R B (1993) Cotton insect losses 1992. *Proc 1993 Beltwise Cotton Prod Res Conference*. pp 655-60. Memphis: National Cotton Council.

Huffakar C B and Lacewell R D (1978) *Biological and Integrated Insect Control with Some Examples of Economic and Energy Savings*. Mimeo, Division of Biological Control University of California Berkeley.

Huffakar C B and Messenger P S (eds) (1976) *Theory and Practice of Biological Control*. pp 750-788. Academic Press, New York.

Huffakar C B and Smith R F (1972) The IBP programme on the strategies and tactics of pest management. *Proc Tall Timbers Conf Ecol Animal Control Habitat Manage*. pp 219-36.

Hugar L B, Veerapur M S and Patil B V (2001) Adoption of integrated Pest Management technology in cotton production in Raichur district of Karnataka state. *J Cotton Res Dev* **15**: 199-201.

Hulme D (1997) *Impact Assessment Methodologies for Micro Finance: A Review*. Manchester, UK.

Hung T Q and Nhat P T (1996) *Vietnam: National IPM Programme-A Country Brief*. Programme Advisory Committee Meeting, FAO-ICP for IPM in Asia, 06-09 February 1996, Hyderabad, India.

Johl S S (1997) Farm sector reorganization for sustainable growth and optimal productivity: A view point. *Proc Third Agricultural Science Congress*. Vol 1. pp 291-300. National Academy of Agricultural Sciences, Punjab Agricultural University, Ludhiana.

Jones D P (1973) Agricultural entomology. In: Smith RF, Mittler TE and Smith CN (eds) *History of Entomology*. pp 307-32. Palo Alto, CA: Annual Review Inc.

Joyce R J V and Roberts (1959) Recent progress in entomological research in the Sudan Gezira, Emp. *Cotton Gr Rev* **36**: 3.

Juliana CS, Annamali R and Samasudran S (1991) Adoption of integrated pest management. *Indian J Ext Edu* **27**: 23-27.

Kalpan A (1994) *The Conduct of Inquiry*. Chandler, San Francisco, USA.

Kamp K, Gregory R and Chowhan G (1993) Fish cutting pesticide use. *ILEIA Newsl* **9**: 22-33.

Kenmore P E (1991) *How Rice Farmers Clean up the Environment, Conserve Biodiversity, Raise More Food, Make Higher Profits. Indonesia's IPM-A Model for Asia.* FAO, Manila.

Kenmore P E (1997) A prerspective on IPM. *LEISA Newsl* **13**: 8-9.

Kenmore P E, Gallangher K D and Ooi P A C (1995) Empowering farmers: experiences with integrated pest management. *Entwicklung and Landlicher Raum* **1/95**: 27-28.

Kenmore P E, Litsinger J A, Bandong J P, Satinago A C and Salac M M (1987) Philippines rice farmers and insecticides: Thirty years of growing dependency and new options for change. In: Tait J and Napompeth B (eds) *Management of Pest and Pesticides: Farmers' Perception and Practices.* pp 98-108. Boulder: West view Press.

Kerlinger F N (1973) *Foundations of Behavioral Research.* Holt Rinehart and Winston Inc, USA.

Kilvin J E and Fliegal F C (1967) Differential perception of innovations and rate of adoption. *Rur Socia* **32**: 78-91.

Kiss A and Meerman F (1991) *Integrated Pest Management in African Agriculture.* World Bank Technical Paper 142. African Technical Department Series World Bank, Washington D.C.

Klassen W (1975) *Pest Management Organization and Resources for Implementation.* (Original not seen. Cited by Kogan M. 1998. *Annu Rev Entomol* **43**: 243-270).

Kogan M (1998) Integrated pest management: Historical perspective and contemporary developments. *Annu Rev Entomol* **43**: 243-70.

Kogan M and Bajwa W I (1999) Integrated pest management: A global reality? *An Soc Entomol Brasil* **28**: 1-25.

Kosloski K (2000) Evaluation research. *Encyclopedia of Sociology 1.* pp 861-869. Macmillan References, New York, USA.

Kovalenkov V G and Aleshev V A (1977) Cotton integrated pest management in Tadjikstan In: Sugonyaev E S and Semyanov (eds) *Biocenological Basis of Cotton Integrated Pest Management.* pp 41-45. Zool Inst Acad Sci Leningard (In Russian). (Original not seen. Cited by Sugonyaev E S .1994. Cotton pest management: Part 5–A Commonwealth of Independent State perspective. *Annu Rev Entomol* **39**: 579-92).

Kranthi K R, Jadhav D R, Wanjari R, Kranthi S and Russel D A (2001 a) Pyrethroid resistance and mechanisms in field strains of *Helicovrpa armigera* Hubner (Lepidoptera: Noctuidae). *J Econ Entomol* **94**: 253-63.

Kranthi K R, Russel DA, Wanjari R, Manoj K, Munje S, Lavhe N, Armes N (2001 b) In-season changes in resistance to insecticides in *Helicoverpa armigera* (Lepidoptera: Noctuidae) in India. *J Econ Entomol* **95**: 135-142.

Lacewell R D and Taylor C R (1978) Economic implication of alternative boll weevil control strategies. *The Boll Weevil Management Strategies, Southern Cooperative Series Bulletin* 288.

Lacewell R D and Taylor C R (1980). Benefit-cost analysis of Integrated Pest Management programs. *Proc of Seminar and Workshop*. pp 283-302. CICP- USAID.

Lacewell R D, Casey J E and Frisbie R E (1974) *An Evaluation of Integrated Cotton Pest Management Programs in Texas: 1964-74*. Texas A & M University, Department of Agricultural Economics: 77-104.

Lal O P (ed) (1997) *Recent Advances in Entomology*, APC Publications Pvt. Ltd, New Delhi.392pp.

Lavekar RC, Sharma O P, Murthy K S, and Puri S N (2001) Integrated pest management in dry land cotton: A case study of Ashta (MS). *J Cotton Res Dev* **15**: 72-78.

Lazersfeld P F and Rosenburg M (1955) The Language of Social Research. (Original not seen. Cited by Shadish W R, Cook T D and Laviton L C.1995. *Foundation of Programme Evaluation*. Sage, Newburg Park, USA).

Lewin K (1948) Resolving Social conflicts: selected papers on group dynamics. (Original not seen. Cited by Shadish W R, Cook T D and Laviton L C.1995. *Foundation of Programme Evaluation*. Sage, Newburg Park, USA).

Lin B N, Vandeman A, Fernanadez C J and Jans S (1994) IPM: How far have we come? *Agric Outlook* No. Ao-207: 24-28.

Lin N (1976) *Foundation of Social Research*. pp 15-41. McGraw-Hill Book Company New Delhi, India.

Litsinger J A (1994) Cultural, mechanical and physical control of rice insects. In: Heinriches EA (ed) *Biology and Management of Rice Insects*. pp 549-584. Wiley Eastern Limited, New Delhi.

Luttrell R G (1994) Cotton pest management: Part 2–A US perspective. *Annu Rev Entomol* **39**: 527-42.

Luttrell R G, Fitt G P, Ramalho F S and Sugonyaev E S (1994) Cotton pest management: Part 1- A worldwide perspective. *Annu Rev Entomol* **39**: 517-26.

Madaus G F, Scriven M and Stufflebeam D L (1991) *Evaluation Models*. Kluwe-Nijhoff Publishing, USA.

Malone S, Herbert D A Jr, Pheasant S (2004) Determining adoption of integrated pest management practices by grains farmers in Virginia. *J Ext* **42**: 1-7.

Mangan M S (1997) Chinese rice farmers: the new IPM experts. *LEISA Newsl* **13**: 14-15.

Matteson P C (1992) 'Farmer First' for establishing IPM. *Bulletin of Entomol Research* **82**: 293-96.

Matteson P C, Gallagher K D and Kenmore P E (1994) Extension of integrated pest management for pant hoppers in Asian irrigated rice: Empowering the user. In: Denno R F and Perfect T J (eds) *Ecology and Management of Plant hoppers*. Chapman and Hall, London.

McNamara K T, Wetzstsesin M E and Douse G K (1991) Factors affecting peanut producer adoption of integrated pest management. *Rev Agric Econ* **13**: 129-139.

Medina C P (1987) Pest control practices and pesticide perception of vegetable farmers in Loo Valley, Benguet, Philippines. In: Tait J and Nampopeth B (eds) *Management of Pests and Pesticides: Farmer's Perception and Practices*. pp 150-157. Boluder West View Press.

Meerman F and van Huis A (1997) Can we make IPM work for resource-poor farmers in sub-Saharan Africa. *International J Pest Magmt* **43**: 313-320.

Merchant M E and Teetas G L (1994) Perception of Texas farmers and pest management advisors on integrated pest management of sorghum insect pests. *South Western Entomol* **19**: 237-248.

Mier C (1995) Training for change: Sowing IPM with small holder farmer in Central America. *MS Interim Report on Ph.D Field Work*. Imperial College, University of London.

Murray D L (1994) *Cultivating Crisis: The Human Cost of Pesticides in Latin America*.Austin Tx: University of Texas Press.

Musser W N, Wetzstsin M E, Reece S Y, Varca P E, Edwards D M, Douce G K (1986) Beliefs of farmers and adoption of integrated pest management. *Agric Econ Res* **38**: 34-44.

Neuman W L (2000) *Social Research Methods-Qualitative and Quantitative Approaches* (Fourth edition). Allyn and Bacon, Boston, USA, USA.

Niyazov O D (1992) Ecological principles of cotton protection. *Proc TBSR Acad Sci Ser Biol Sci* **2**: 3-13 (Original not seen. Cited by Sugonyaev E S .1994. Cotton pest management: Part 5–A Commonwealth of Independent State perspective. *Annu Rev Entomol* **39**: 579-92).

Nody I and Chhay N (1996) *Kingdom of Combodia: National IPM Programme: A Country Brief*. Programme Advisory Committee Meeting FAO-ICP for IPM in Asia, 06-09 February, 1996, Hydrabad, India.

Norris R F, Caswell-Chen E P and Kogan M (2002) *Concept in Integrated Pest Management*. Prentice-Hall of India Private Ltd, New Delhi.

Norton G A (1993) Agricultural development paths management: A pragmatic view of sustainability. In: Chadwick D J and Marsh J (eds) *Crop Protection and sustainable Agriculture*. pp 100-115. John Wiley and Sons Chicehster.

NRC (1989) *Alternative Agriculture*. National Research Council. National Academy Press, Washington DC.

Ooi P A C (1998) Beyond the farmer field school: IPM and empowerment in Indonesia. *Gatekeeper Series, IIED-VCL* Environmental Economics Centre Development 78:116.

Ooi P A C (2003) IPM in cotton: lessons from rice IPM. In: Mew T W, Brar D S, Peng S, Dawe D and Hardy B (eds) *Rice Science: Innovations and Impact for Livelihood*. International Rice Research Institute, Philippines.

Ooi P A C, Lim G S, Ho T H, Manalo P L and Waage J (eds) (1992) *Integrated Pest Management in the Asia Pacific Region*, CAB International Bank, Kuala Lumpur.

Ortiz O (2001) *Workshop on Assessing the Impact of Participatory Research and Extension Methods in IPM-Reports of Working Groups* (first draft), Nairobi 15-17 March, 2001, Nairobi.

Oudejans J H (1991) *Agro-Pesticides: Properties and Functions in Integrated Crop Protection*. United Nations Economic and Social Commission for Asia and Pacific, Bangkok.

Owen J M (1993) *Programme Evaluation, Forms and Approaches*. Allen and Unwin, St. Leonard, NSW, Australia.

PAU (1979) *Package of Practices for Crops of Punjab–Kharif*. Directorate of Extension Education, Punjab Agricultural University, Ludhiana.

PAU (1991) *Package of Practices for Crops of Punjab-Kharif*. Punjab Agricultural University, Ludhiana.

PAU (2004 and 2005) *Package of Practices for Crops of Punjab-Kharif*. Punjab Agricultural University, Ludhiana.

Peshin R (2002) Economic benefits of pest management. In: Pimental D (ed) *Encyclopedia of Pest Management*. pp 224-227. Marcel Deker Inc, New York.

Peshin R and Kalra R (1998) Integrated pest management at farmer's level. *Man and Dev* **22**: 137-141.

Peshin R and Kalra R (2000a) Farmers field schools: Facilitating participatory learning . *J Ext Edu*, **11**: 2932-2936.

Peshin R and Kalra R (2000b) *Integrated Pest Management: Adoption and its Impact on Agriculture*. Classical Publishing Company, New Delhi.

Peshin R and Kalra R (2002) Constraints in the adoption of IPM practices; A case study of Ludhiana district, Punjab. *Indian J Ecol* **29**: 49-53.

Peshin R, Kalra R and Kaul V K (1997) Impact of training under farmer field school on the adoption of IPM practices in rice crop. *Proc Third Agricultural Science Congress* **2**: pp 365 PAU, Ludhiana, India.

Petheram R J (1998) *Review of Evaluation in Agricultural Extension*. Rural Industries Research and Development Corporation, Kingston. **http://www.rirdc.gov.au**.

Pickett A D and Patterson N A (1953) The influence of spray programme on the fauna of apple orchards in Nova Scotia IV: A review. *Can Entomol* **85**: 472-87.

Pickett A D, Putman W L and Le Roux E J (1958) Progress in harmonizing biological and chemical control of orchard pests in Eastern Canada. *Proc Tenth International Congress of Entomology* **3**: 169-174.

Pimental C R M and Banderia C T (1981) Beneficious obtidos pelo productor atraves do manejo de pragas na cultura algodoeira na regiao centro-sul. *EMBRAPA/ CNPA Comun Tec* **18**: 13.

Pincus J (1991) Farmer field school survey: Impact of IPM training on farmers' pest control behaviour .*IPM National Programme*. pp 103-110. Ji Ki Mangunsarkoro S, Jakarta.

Pingali V (2004) *State of the Indian Farmer – A Millennium Study: Input Management*. pp 197-242. Ministry of Agriculture, Government of India, New Delhi.

Pontius J, Dilts R, Bartlelt A (2002) *From Farmer Field School to Community IPM-Ten Years of IPM Training in Asia*. FAO Community IPM Programme Jakarta, Indonesia. **http://www.fao.org**

Posavac E J and Carey R G (1997) *Programme Evaluation, Methods and Case Studies*. Prentice Hall, Englewood Cliff, New Jersey.

Poswal A and Williamson S (1998) Off the 'treadmill': Cotton IPM in Pakistan. *Pesticides News* **40**: 12-13.

Pradhan S (1964) Assessment of losses caused by insect pests of crops and estimation of insect population. In: *Entomology in India*. pp 17-58. Entomological Society of India, New Delhi.

Pretty J N (1994) Alternative systems of inquiry for sustainable agriculture. *IDS Bulletin*. University of Sussex **25**: 37-48.

Pretty J N (1995) *Regenerating Agriculture: Polices and Practices for Sustainability and Self-Reliance*. Vikas Publishing House Pvt Ltd, New Delhi.

Provus M M (1971) *Discrepancy Evaluation/ or Education Programme, Improvement and Assessment* CA: Mc Cutchan, Berkely.

Puri S N (1995) Present Status of IPM in India. *National Seminar in Integrated Pest Management*. Dec 29-30, Nagpur, Maharashtra.

Qadeer G A and Tomar R S (1993) Rice IPM demonstration in Haryana: A success story. *Pl Prot Bull* **45**: 21-23.

Ragunathan V (1995) Integrated pest management: Eco sound plant protection. *The Hindu: Survey of Indian Agriculture*, Chennai, India: 154-155.

Ragunathan V (2005) Ecofriendly pathways. *The Hindu Survey of Indian Agriculture*, Chennai, India: 160-164.

Ramalho F S (1994) Cotton pest management: Part 4–A Brazilian perspective. *Annu Rev Entomol* **39**: 563-78.

Ramasamy S S (1994) Price effect demand for pesticides in cotton: A micro analysis . *Indian J Agric Mar* **8**: 61-71.

Ridgway R L, Lloyd E P and Cross W H (eds) (1983) *Cotton Insect Management with Special References to the Boll Weevil*. USDA Agric Research Sevice Handbook, pp 591.

Rogers E M (1983 and 1995) *Diffusion of Innovation*. Free Press, New York.

Rola A C and Baril T A (1997) Making rice farmers better decision makers via the farmer field school. *SEMEIO Update* **5**: 10-12.

Rolling N (1988) *Extension Science: Information Systems in Agricultural Development*. Cambridge University Press, Cambridge, UK.

Romer P M (1990) Endogenous technical change. *J Pol Econ* **98**: pt **2**: 571-5102.

Rossi P H and Freeman H E (1993) *Evaluation* (fifth ed). pp 5 Sage Newbury Park, California, USA.

Rossi P H, Freeman H E and Write S R (1979) *Evaluation: A systematic Approach.* Sage Publishers, Bawerly Hills, London.

Rossing W A H, Schans J and Zadoks J C (1985) Het project Epipe, project analysis in de Gawasbescherming. *Land Bouwkunding Tijdscrift* **23**: 455-473. (Original not seen. Cited by van de Fliert E. 1993. *Integrated Pest Management: Farmer Field School Generate Sustainable Practices.* Wageningen Agricultural University Papers 93.3).

Rubia E G, Lazaro A A, Heong K L, Diah, Nurhasyim and Norton G A (1996) *Crop Prot* **15**: 327-333.

Russell D (2004) Integrated pest management in less developed countries. In: Horowitz AR and Ishaaya I (eds) *Insect Pest Management: Field and Protection Crops.* pp 141-208. Springer, New Delhi.

Russell D, Kranthi K R, Mayee C D, Banerje S K and Raj S (2003) Area-wide management of insecticide resistant pests of cotton in India. *Proc The World Cotton Conference-3*, 9-14 March 2003, Cape Town South Africa.

Rutman L (1984) *Evaluation Research Methods.* pp 10. Sage New bury park, California.

Salkin M S, Eidman V R and Massey W B Jr (1976) *An Economic Analysis of Some Alternative Pest Control Strategies for Grain Sorghum in the Oklahoma Panhandle.* Agricultural Experiment Station B –722, Oklahoma State University.

Sattar M A and Shafiquddin K (1996) *Bangladesh: National IPM Programme: A Country Brief.* Programme Advisory Committee Meeting, FAO-ICP programme in Asia, 06-09 February, 1996, Hydrabad, India.

Saunders M, Lewis P and Thronhill A (2004) *Research Methods for Business Students.* Pearson Education Pvt Ltd, Singapore.

Scriven M (1967) The methodology of evaluation. In: Stake R E (ed) *Curriculum Evaluation.* R and Mc Nally, Chicaogo USA.

Scriven M (1972) Pros and cons about the goal free evaluation. *Evaluation Comment* **3**:1-5.

Scriven M (1980) *The logic of Evaluation.* Edgepress Inverness California (Original not seen. Cited by van de Fliert 1993. *Integrated Pest Management: Farmer Field Schools Gebnerate Sustainable Practices.* Wageningen Agricultural University Papers 93.3).

Scriven M (1991) *Evaluation Thesaurus* (4th ed). Sage, New-bury Park, California, USA

Sechrest L, Stewart M, Stickle T R and Sidani S (1996) *Effective and Persuasive Case Studies.* Toolkit commissioned by the Evaluation Centre at Human Service Research Institute, Cambridge, MA.

Seganmullar A and Hewson RT (2000) Global implementation of ICM in cotton. *Proc the Brighton Crop Conference 2000: Pest and Disease.* pp 193-198. British Crop Protection Council, Farnham, surrey, UK.

Sen A and Bhatia M S (2004) *State of the Indian Farmer: A Millennium Study-Cost of Cultivation and Farm Income* Vol 14. Ministry of Agriculture, Government of India, New Delhi.

Shadish W R (1993) Critical multiplism: A research study and its attendant tactics. In: Sechrest L (ed) *Programme Evaluation: A Pluralistic Enterprise*. Jossey-Bass, San Francisco USA.

Shadish W R, Cook T D and Leviton L C (1995) *Foundations of Program Evaluation*. Sage, Newbury Park, California, USA.

Shadish W R, Thomas D C and Laura C L (1991) *Foundations of Program Evaluation: Theories and Practice*. 529 pp. Sage, Newbury Park, California, USA.

Shetty P K (2004) Social-ecological implication of pesticide use in India. *Econ and Pol Weekly* **39**: 5261-5267.

Simwat G S (1994) Modern concepts in insect pest management in cotton. In Dhaliwal GS and Arora R (eds) *Trends in Agricultural Insect Pest Management*. pp 186-237. Commonwealth Publisher, New Delhi.

Singh D (2002) Adoption of recommended integrated pest management (IPM) practices by cotton growers of Bathinda district (Punjab). *M.Sc. Thesis Punjab Agricultural University, Ludhinana, India*.

Singh P and Ramasundaram P (2005) Cotton: big boost for textile industry. *The Hindu Survey of Indian Agriculture*: 101-109.

Singh Raghbir and Peshin R (2000) Operationalizing the definitions for extent and level of adoption. In: Peshin R and Kalra R. *Integrated Pest Management: Adoption and its Impact on Agriculture*. p 26. Classical Publishing Company, New Delhi.

Singh Ravindra (1975) On optimum stratification for proportional allocation. *Sankhya* **37**: 109-115.

Singhal V (2003) *Indian Agriculture 2003*. pp 310-334. Indian Economic Data Research Centre, New Delhi.

Smith M F (1989) *Evaluability Assessment: A Practical Approach, Evaluation in Education and Human Services*. Kluwer Academic, Boston, USA.

Smith R F and Allen W W (1954) Insect control and the balance of nature. *Scientific American* **190**: 38-92.

Smith R F, Mittler T E and Smith C N (1973) (eds) *History of Entomology*. Palo Alto, CA: Annual Review Inc.

Sorensen A A (1993) *Regional Producer Workshop: Constraints to the Adoption of Integrated pest Management*. Office of Pesticide Programmes, US, EPA, 51pp.

Sorensen A A (1994) *Proc National Integrated Pest Management Forum Arlington*. pp 1–86. Am Farmland Trust.

Sprott J M, Lacewell R D, Niles G A, Walker J K and Gannaway J R (1976) *Agronomic, Economic, Energy* and *Environmental Implication of Short Season, Narrow Row Cotton Production*. Texas Agricultural Experiment Station, Texas, USA.

Stake R E (1973) Programme evaluation: Particularly responsive evaluation. In Madaus GF, Scriven M S and Stufflebeam D L (eds) *Evaluation Models*. Kluwer-Nijhoff, Boston, USA.

Stake R E (1995) *The Art of Case Study Research*. Sage Publications, Newbury Park, California, USA.

Stern V M, Smith R F, van de Bosh R and Hagan K S (1959) The integrated control concepts. *Hilgradia* **29**: 81-101.

Sugonyaev E S (1994) Cotton pest management: Part 5-A Commonwealth of Independent States Perspective. *Annu Rev Entomol* **39**: 579-92.

Swaminathan M S (1975) ICAR, Operational Research Projects, purpose and approach. *Indian Farming*, August 1975.

Swezy S L, Murray D L and Daxal R G (1986) Nicaragua's revolution in pesticide policy. *Environment* **28**: 6-9.

Thomas J K, Ladewig H and Macintosh W A (1990) The adoption of integrated pest management practices among Texas cotton growers. *Rur Sociology* **55**: 365-410.

Tyler R W (1967) Changing the concept of educational evaluation. In Stake RE (ed) *Perspectives of Curriculum Evaluation*. Rand Mc Nally, New York, USA.

Uphoff N (1989) *A Field Methodology for Participatory Self-Evaluation of PPP Group and Inter-group Association Performance*. People's Participation Programme of the UN Food and Agriculture Organization, Ithaca, New York.

Uphoff N (1992) *Participatory Evaluation of Rural Development*. IFAD Monitoring and Evaluation Division for Monitoring and Evaluation Panel, New York.

USDA (2006) United States Department of Agriculture. Foreign Agricultural Services. http://www.icac.org.

van de Fliert E (1993) *Integrated Pest Management: Farmer Field School Generate Sustainable Practices*. Wageningen Agricultural University Papers 93.3

van de Fliert E (1998) Integrated pest management: Springbord to sustainable agriculture. In Dhaliwal GS and Heinrichs EA (eds) *Critical Issues in Insect Pest Management*. pp 285-304. National Agricultural Technology Information Centre, Ludhiana, India.

Vanyants G M (1991) Organisation and economic base of cotton integrated pest managemnt in Tadjikistan. *CSC Thesis St. Petersburg–Pushkin*. (Original not seen. Cited by Sugonyal E S .1994. Cotton pest management: Part 5-A Commonwealth of Independent States perspective. *Annu Rev Entomol* **39**. 579-92).

Wahab S (1997) Biotechnological approaches for the management of pests, diseases and weeds in agriculture-country status. *Pestic World* Sept-Oct: 13-21.

Waibel H (1999) Policy perspective of IPM evaluation. In Waible H, Flesiher G, Kenmore PE and Feder G (eds) *Evaluation of IPM Programmes- Concepts and Methodologies*. pp 13-20. Pesticide Policy Project Publication Series 8. Hanover.

Waibel H, Fleischer G, Kenmore P E, and Feder G (eds) (1999) *Evaluation of IPM Programs–Concepts and Methodologies*. pp 43-61. Publication series no 8 of the Pesticide Policy Project, Institute of Economics in Horticulture, Hannover.

Wholey J S (1983) *Evaluation and Effective Public Management*. Little Brown, Boston, USA.

Williams I H (2004) Advances in insect pest management of oilseed rape in Europe. In: Horowitz A R and Ishaaya I (eds) *Insect Pest Management: Fields and Protected Crops*. pp 181-208. Springer International, New Delhi.

Williamson S (1999) *Report on Organic Cotton and Farmer participatory Training in Latin America*. CABI Bioscience UK.

Wilson L J, Mensak R K and Fitt G P (2004) Implementing integrated pest management in Australian cotton. In: Horowitz AR and Ishaaya I (eds) *Insect Pest Management – Field and Protected Crops*. pp 97-140. Springer, New Delhi.

Winarto Y T (1992) Farmers' agro ecological knowledge construction: the case of integrated pest management among rice farmers in north cost West Java. *Paper presented at IIED/ Ds Beyond Farmer First: Rural People's Knowledge Agricultural Research and Extension Practice Workshop*. IDs University of Sussex, 27-29 Oct, 1992.

Winarto Y T (1995) State intervention and farmers creativity: Integrated pest management among rice farmers in Subang, West Java. *Agric and Human Values* **12**: 47-57.

Wiradmadya R and Kusmayadi A (1996) *Indonesia: National IPM Programme. A Country Brief*. Programme Advisory Committee Meeting FAO ICP for IPM in Asia, Hyderabad India.

Worthen B R, Sanders J R and Fitzpatrick J L (1997) *Programme Evaluation: Alternative Approaches and Practical Guidelines*. Longman, New York.

Yogfan P and Diyun L (1996) *PR China: National IPM Programme-A Country Brief*. Programme Advisory Committee Meeting, FAO, ICPP for IPM in Asia, 06-09 February, Hyderabad, India.

Zadoks J C (1985) On the conceptual basis of crop loss assessment: The threshold theory. *Annu Rev Phytopathol*. **23**: 455-473.

Zadoks J C (1989a) Does IPM, give adequate attention to disease control, in particular at farmer level? *FAO Pl Prot Bull* **372**: 145-149.

Zadoks J C (1989b). EPIPRE: a computer based decision support system for pest and disease control in wheat: Its development and implementation Europe. *Pl Disease Epidemiology* **2**: 3-39.

Zaltman G, Duncan R and Holbek J (1973) *Innovations and Organizations*. John Wiley and Sons, New York.

Zavaleta L R and Ruesink W R (1980) *Expected Benefits from Non-Chemical Methods of Alfalfa Weevil Control*. Mimeo, Section of Economic Entomology, University of Illinois.

INDEX

T